# Prologue

It has been a long time, almost 30 years, since I met and participated with ICO. NoAH has been published 9 times already. Every 2 – 3 years, we have seen one of their most prestigious package design annual book worldwide.

Why is it so often? Graphic design is like the fashion industry. Every decade, every year, every season people's expectations change. Based on the market, economy, technology, ecology and so forth. Especially package design for retail stores, supermarkets and the food industry. Almost every year, every season, the trends change. We have to study, learn and create new things everyday.

If you look at the graphic design industry 30 years ago, there is no computer in any graphic designer's hand. There are big differences with the computer and without the computer to designing any package graphics. Now people are concerned about the environment and being "green". This environmental awareness influences graphic design.

Hopefully NoAH 9 Package Design Annual, showing all kinds of design from all over the world helps to teach what all the graphic designers are doing now.

Even myself, being in this industry over 30 years, still studies and learns everyday to stay in touch with current waves and trends.

Yashi Okita

*aloof design uk*

**HUMBER LIBRARIES LAKESHORE CAMPUS**
3199 Lakeshore Blvd West
TORONTO, ON.   M8V 1K8

**Brands. Packaging. Needs. Wants. Beauty. Choices.**

Hundreds of cereal boxes to choose from. Dozens of ice cream flavors to choose from -all in one store. 'New, natural, family size, individual servings, extra bonus size, sugar-free, generic brands, two-for-one, six-pack, on and on'. What a world of shopping we live in.

Packaging always grabs my attention. Packaging helps me notice if the product is one of the more prestigious brands, something new and different, or a new line extension of an existing brand. I am continually sampling new products that I have never noticed or tried before. As such, package designers battle for shelf space with the creation of excitingly expressive packages that utilize vivid colors, bold lines, and strong typography.

The challenge for designers also continues to be the building of customer relationships. This is more than just packaging. A well-designed package can help make the purchasing decision at the point of sale and may also enhance customer loyalty. Some of my favorite trends include 'the Green Look,' which shows itself visually through the use of obviously earthy materials or the use of simple containers that are made up of fewer materials. Smaller would be great as well.

It is hard to not engage in a conversation about packaging without talking about 'green and eco friendly'. In my opinion, American products come in way too much packaging. Products are encased in layers and layers of plastic, from the hard plastic inside the box to the clear plastic that seals the box. I would personally challenge package designers to compete for the most minimal, recyclable, non plastic packaging with the most attractive designs. Packaging must use recycled materials, few ( if any ) chemicals, and be inexpensive to produce. I shop almost daily and try to do my part. I do this with simple things like skipping plastic bags in favor of canvas bags, walking and driving less whenever possible, buying items that don't 'over package', and supporting companies that use recyclable materials.

Many companies are successfully using new ways to simplify the look and feel of packages while making them more noticeable and user friendly. It seems that designers all over the world are continuing to find creative new ways to be less wasteful. These designers are striving to do their bit to keep the planet more sustainable by seeking ways to go in the 'green' direction.

Unless we recycle and reuse, packaging will continue to seem to many only wasteful. However, delivering products through packages, then onto consumers, remains a powerful and essential part of the design and marketing process. The ongoing challenge for package designers is to maintain their creative edge while being both economically and environmentally responsible. If we can continue to make further advances of innovative design and technology, then I believe we can meet the challenge of being more 'expressive' and less 'excessive'.

Mike Quon

Principal & Creative Director
Designation Inc. Design as a function of prosperity

**Packaging**

In this commercially competitive global market packaging is a vital part of the decision-making processes for consumer citizens to select and purchase certain products among a multitude of products that are available in marketplaces.
In addition, packaging takes on added importance in a computer age where dissemination of information is instantaneous and omnipresent for consumer scrutiny.

A well-packaged product with attractive and soothing artworks will have a definite advantage in luring consumer citizens to select and purchase that particular product over other similarly situated products.
For an instance, it is believed that Barrack Obama won the U.S. 2008 presidential election over John McCain due largely to his appearance, his demeanor, and his style of delivering speeches in public arenas. He was better packaged than John McCain.
And who is doing the yeoman's work in this indispensable aspect of creating and designing product packaging that wins over the consumer's hearts and minds? They are commercial artists whose creative talents and artworks come from around the world.

I have known and befriended Norio Mochizuki, who is the founder of International Creators' Association that publishes noAH, among other publications. He is a pioneer in organizing a world-wide organization of commercial artists for the purpose of introducing their artworks in the world market places.
His contributions toward the commercial art field throughout the world for the past 40 some odd years are outstanding and immeasurable.

The computer age has produced many unpredictable consequences in commercial market places. As trends in the packaging industry have changed over the years, Norio Mochizuki has continued to report them to you in ICO's noAH series.
My hat is off to him for all his years of hard work. Congratulations to Mr. Norio Mochizuki. We look forward to keeping informed about the latest trends of global packaging in future noAH publications.

Koichi Yanagizawa / Lawyer
in Los Angeles, California

**Design as a function of prosperity**

We have heard a lot of global design, just as much as Brazilian or French or Korean Design. Each country or culture tries to use design as a tool of identity and differentiation. It is a world where there is a strong force is beginning to merge together all communications.

Globalization is definitely a strong development vector powered by costs and scale measures. This adds a strong push to world design. Nevertheless, the real driving force behind this phenomenon is the unstoppable pursuit of comfort by the human being. We want better, faster, cleaner and trendier. And after we have purchased it we expect to position ourselves to the world based on what we buy. Simple as that. If for a good price so much better, but buying expensive also make a statement.

I am Brazilian to the bone, and the base of my thinking either comes from my country's peculiar history and way of life, or despite of it. Brazil's Carnaval - I assume - is our most well known identity contribution to the world. And Carnaval (not Carnival please) is a very good example of what design is all about. This country has a majority of people living by or in the poverty line, but even though they will work the year out to pay for their costumes to run the Avenue and get a deserved 4 hour of illusion. A well known director of Samba School's said once that the poor love luxury; misery is the matter of intellectuals.

**And prosperity brings a greater demand on well being goods and design.**

For being a design company manager I have realized that, in my particular region, design is definitely becoming more necessary as consumers have more to spend and more access to goods. Our own company history reflects that. We were founded in Brazil and later opened an office in Chile, then Argentina, and later in Mexico. Design work now is in real demand in Colombia, a country that for the last 8 years has shown a reintegration to the world's economy, and Peru, a country that has grown 10% in the last year.

**Good for designers but even better for consumers.**

Economic prosperity is a strong vector for the relevance of design. Prosperity adds choice to where there was not, and products start to be chosen for factors other than availability only. People will think whether the new option is tastier, better or worth a try, and buy it and change the dynamics of the market.

To this effect companies also seek for more efficient ways to get the better products to the consumer, trying to grab a share of the best design in the world. For example: now we design works of BMW in the US, certainly set in respect to BMW's largest market, so as to follow and better develop its products to the taste of its consumer. And we see design centers open in every country like the UK, France, Italy, Japan and Brazil itself, and many, many others. The impact of these initiatives is major, and companies are just beginning to touch the surface of a vast resource of inspiration and ideas that lie there waiting to be brought to market.

We ourselves operate like a spread-out company, being one office with 4 different bases. Our personnel has moved from every county to every other, making us a really global operation with a strong Latin root.

And we see it coming. Despite of the small relative economic size of our region, I have witnessed the growth of its important growing exponentially as it carries a large share of the world's youth, consumers of tomorrow, and an even larger share of global resources, in a still yet untouched and mostly virgin territory.

Sustainability initiatives abound, but with a very different tone. Here we are not desperate to rescue a last tree, but better, we are starting to implement a new development model, were resources and nature are plenty. Man has not yet depleted our region of its sources, and we can preserve it rather than try to rebuild out of ashes.

It may not happen today, but with a strong pace and eyes into the future we will get to development respecting our environment and our people, under a different model of prosperity.

Fernando Muniz-Simas

*DIL BRANDS*
*Brasil Argentina Chile Mexico*

# LEWIS MOBERLY
### ENGLAND

**Our disciplines** : Brand consultancy, brand identity, name generation, interactive design.

**Our clients** : Moet et Chandon, Pernod Ricard, Harrods, Johnson and Johnson, Pepsico, Marks and Spencers, Kao Corporation, Groupe Accor, Tefal, Bollinger, Champagne Pol Roger, International Tennis Federation.

**Our approach** : LM prides itself on a thoughtful approach. We believe that design starts with great detective work. Our first step is to rigorously uncover truths about the brand, its consumers, competitors and the culture in which it lives. This enables us to manage complex brand hierarchy and communication issues.

This approach has led to countless awards, for both creativity and effectiveness. At LM we believe one gets you the other. We are the only consultancy to hold the top awards for creativity, D&AD Gold, and effectiveness, DBA Grand Prix.

**Our philosophy : First win the eye, then the heart, then the mind**
Design is the most enduring expression of a brand and design must work on three levels. First it must catch your attention. Then engage on an emotional level, the first step to building a brand relationship. Finally it must rationally convince you of the brand's promise.

Lewis Moberly
33 Gresse Street
London W1T 1QU
United Kingdom
Tel: +44 (0) 207 5809252
Fax: +44 (0) 207 2551671
Contact: Pierre Boyre
Pierre.boyre@lewismoberly.com
www.lewismoberly.com

Waitrose Ltd, United Kingdom
Range of Mustards
Food
Glass Jar with Die Cut Paper Label
Label Design

Waitrose Ltd, United Kingdom
Range of Fresh Herbs
Food
Flow Wrap Bag
Design and Copywriting

Waitrose Ltd, United Kingdom
Cooks' Ingredients
Food and Drink
Paper Labels
Range Creation, Identity and Copywriting

Waitrose Ltd, United Kingdom
Cooks' Ingredients
Food and Drink
Diverse Materials
Range Creation, Identity and Copywriting

William Grant & Sons, United Kingdom
Glenfiddich
Single Malt Whisky
Metalised Board Canisters
Identity, Label and Canister Design

The Glenmorangie Company (LVMH)
United Kingdom
The Glenmorangie
Single Malt Whisky
Glass Bottle and Rigid Box
Identity, Label and Carton Design

Bodegas Chivite, Spain
Arínzano
Alcoholic Drink – Wine
Glass Bottle, Paper Label, Foil Capsule
Identity and Packaging Design

Champagne Bollinger, France
Bollinger NV Special Cuvée
Alcoholic Drink– Champagne
Glass Bottle and Folding Carton
Identity, Label and Carton Design

Selfridges & Co. United Kingdom
Luxury 'House Brand'
Food and Drink
Diverse Materials
Identity and Range Creation

# CATO PARTNERS
## AUSTRALIA

### Australia

**Melbourne**
Cato Purnell Partners
10 Gipps Street
Collingwood 3066
Victoria
Telephone +61 3 9419 5566
Facsimile +61 3 9419 5166
melbourne@cato.com.au

**Sydney**
Cato Purnell Partners
489 Elizabeth Street
Surry Hills 2010
New South Wales
Telephone +61 2 9318 1111
Facsimile +61 2 9318 1899
info@catosydney.com

**Adelaide**
c/- Einstein Da Vinci
94-96 Fullarton Road
Norwood 5067
South Australia
Telephone +61 8 8362 5211
Facsimile +61 8 8362 8633

**Brisbane**
84 Cavendish Street
Nundah 4012
Queensland
Telephone +61 7 3314 6229

### International

**New Zealand**
Cato Partners NZ Limited
Level 9
175 Victoria Street
Wellington
Telephone +64 4 499 5549
Facsimile +64 4 499 6226
mail@cato.co.nz

**Singapore**
Consulus Cato Partners
23 Hindu Road 209115
Singapore
Telephone +65 6293 9495
Facsimile +65 6293 9485
lawrence@consuluscato.com

**Spain**
Cato Partners Europe
Calle Costa Brava, 13 2
Madrid 28034
Telephone +34 91 735 5689
astalman@catospain.com

**United Arab Emirates**
Cato Partners
Mot2iwalla
The Complete Digital Resource
Dubai 31788
Telephone +971 4 359 6188
Facsimile +971 4 355 2889
motiwalla@cato.com.au

**México**
Cato Saca Partners
Terranova 714-A int.2
Providencia
Guadalajara 44670
Telephone +52 33 3640 6566
Facsimile +52 33 3617 1365
rsaca@catosacapartners.com

**India**
Design Protocol Pvt. Ltd F.
Unit 10, Drego House
Dr Ambedkar Road
Bandra (West)
Mumbai 400-050
Telephone +91 22 2605 1579
Facsimile +91 22 2600 8075
designprotocol@vsnl.net

**Indonesia**
Cato Partners
PT Komunikasia
Jl Danau Terusnan B2 no 85
Bendungan Hilir
Jakarta 10210
Telephone +62 21 251 1295
Facsimile +62 21 570 6639
gunawan@cato.com.au

*In 1970, Cato Partners set out on a journey of exploration and experimentation in the search for a better understanding of the visual and verbal components that identify and differentiate one company, organisation or event from that of its competitors. Today that journey continues. With the accumulated experiences of work in almost every business sector and cross-cultural interaction with companies in 35 countries, that understanding has become clearer and the strategic visual alternatives, opportunities and possibilities have become far greater. The company's development of the concept of Broader Visual Language™ has greatly enhanced the ability for companies to utilise their brands more appropriately in response to business opportunities and maximise every occasion to express the brand's presence to the fullest.*

1 1 Fosters Group
　2 Hubert the Stag
　3 Pinot Noir

2  1  Underground Winemakers
   2  Pinot Grigio
3  2  Violets Moscato
4  2  Cab Merlot
5  1  Fosters Group
   2  T'Gallant Romeo

6   1   Melba Recordings
      2   Tzigane
7   2   Seduction
      3   Album Covers

8  1. Matilda Bay Brewing Company
    2. Beez Neez
    3. Honey Wheat Beer

9  1  Balnarring Vineyard
    2  Pinot Grigio Friulano
10  2  Pinot Noir
11  2  Extremely Funky White

# ESTUDIO IUVARO
## ARGENTINA

Estudio Iuvaro is a packaging design consultancy based in Mendoza, Argentina, at the foothills of The Andes. We've been working in wine communication since 1994. Our job is focused on packaging design for beverages (specially wines) and gourmet products.

Since for small wine cellars wine labels are the only marketing tool they possess to attract consumers, we make it our priority to approach the label design based not only on professionalism, but also on the consideration of the emotions the product has to evoke. Hence, we work cooperatively with our clients taking into consideration their ideas and feelings. Through the development of identity programs for the products we also provide creative and branding solutions for our clients worldwide.

Furthermore, we strongly believe there should be a connection between the label and what the product inspires in the consumers' minds. Label design is not simply dressing the bottle, it implies the understanding of the product in order to design a label that will satisfy the consumers' expectations.

Cecilia Iuvaro, General Director with Silvia Keil, Celia Grezzi and Valeria Aise (who complete the team) we are all graduated from Universidad Nacional de Cuyo, established in a well known region of vineyards, the most important one in Argentina. Hence, designing is a matter of culture and roots for us.

Our aim is to produce distinctive ways of showing how wine can fire the consumers' imagination. Through extensive research, constant innovation and a dose of pleasure we design labels considering that we are both designers and consumers.

1  Jaure Winery | USA
JJ (Jacinto Jaure) | 2006
Packaging design

2  Santa Carolina Winery | Chile
VSC (Icon wine) | 2007
Packaging design

3   Viña Las Perdices Winery | Argentina
    Tinamú (Icon wine) | 2007
    Packaging design

4   Viña Las Perdices Winery | Argentina
    Las perdices Ice wine | 2007
    Packaging design

5   Finca Vistaflores S.A. | Argentina
    Campo Los Andes Wine | 2005
    Packaging design

6   Concha y Toro Winery | Chile
Casillero del Diablo Reserva Privada | 2006
Structural packaging and label design

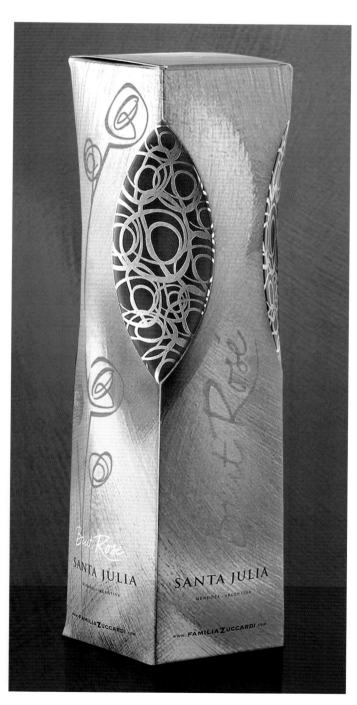

7  Familia Zuccardi Winery | Argentina
   Santa Julia Brut Rosé | 2008
   Gift box and packaging design

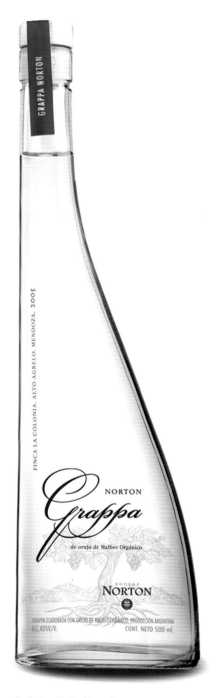

8  Bodega Norton | Argentina
   Grappa | 2008
   Packaging design

9 Familia Zuccardi Winery | Argentina
Santa Julia Magna | 2008
Packaging design

10  Familia Zuccardi Winery | Argentina
    Vida Orgánica Range | 2008
    Packaging design

11  Bodega Norton | Argentina
    Norton Range Labels | 2001
    Packaging design

# VILLEGERSUMMERSDESIGN
## UNITED KINGDOM

Founders Angela Summers and Vincent Villéger

Vincent Villéger and Angela Summers established villégersummersdesign nearly a decade ago. The business was born out a fundamental desire to deliver quality bespoke design, whilst also providing an excellent service.

This year, villégersummersdesign were awarded a Gold Pentaward for their design of the Givenchy "Ange ou Démon" refillable bottle.

A personal, flexible and pro-active approach has enabled villégersummersdesign to grow and develop over the years, winning the trust of many major brands such as Yves Saint Laurent, Givenchy, Issey Miyake, Boucheron, Van Cleef & Arpels and the Body Shop.

Internationally minded, we are happy to communicate in French as well as English, and regularly travel between London and Paris for business.

Our ability to understand a company's needs, identity, and objectives has resulted in successful designs and sustained relationships with many of our clients.

Our expertise in the luxury sector can be applied to bottle design, product design, packaging design, and promotional design (boxsets, gift with purchase, etc).

Starting from a sketchbook of ideas, we will work closely with your product development team to generate and develop concepts.

Your feedback is welcome at every stage of the process, ensuring a result which will match your requirements and exceed your expectations.

Here is what some of our clients have to say about us:

"If your project deadline is far enough, if your budget is large enough, if you are not really looking for fresh and out-of-the-box ideas, then you could probably pass on this agency. My reality was more brutal and working with villegersummersdesign always guaranteed soft landing projects. Internationally minded, always interacting with a smile. Simply talented."

NV, product manager, Issey Miyake Perfumes

"The objective for Givenchyman was to create a range with a design that was elegant and simple, yet impactful.
The design of the Givenchyman range is a global success, from the primary packs down to the secondary and sampling packs.
It was given an identity that is genuinely coherent with the Givenchy brand and with the specific expectations of the men's cosmetics sector.
The launch of the range was an immediate success. Now, this success continues and will be backed-up by the launch of new additions to the Givenchyman range."

AS, International Marketing, Givenchy Perfumes

**www.villegersummersdesign.com**

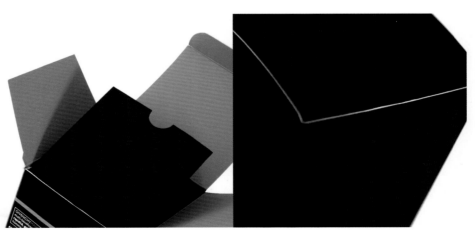

Givenchyman secondary packaging and brand identity

Givenchyman primary packaging and brand identity

Givenchy Ange ou Démon refillable perfume bottle - structural design

Givenchy Ange ou Démon refillable perfume bottle chain design - structural design

Yves St Laurent Opium purse spray perfume bottle - structural design

Van Cleef & Arpels Christmas box set - promotional packaging design

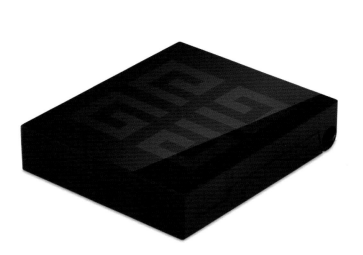

Givenchy Matissime compact face powder - product design

# ANNETTE SCARFE DESIGN
## SPAIN

## ANNETTE SCARFE DESIGN
### BRAND DEVELOPMENT

Annette Scarfe Design is a unique Brand Development and Design company, which specialises in the International Wines and Spirits sector. The company combines people with many years experience and a dedicated and unique approach to brand development..

Our European office in Spain, world wide representation and a bespoke client intranet, enables us to provide premium design services to all our International clients.

For more information contact Roland Antony.
Annette Scarfe Design, Apartado 731, Valls 43800 Tarragona Spain. Telephone (34) 977 050 120. Fax (34) 977 050 121
email design@annette-scarfe.com
www.annette-scarfe.com

1. 1. Direct Wines - United Kingdom
   2. Tenca Tree
   3. Chilean Wine
   4. Self adhesive label
   5. Shiraz

2. 1. Renzo Masi - Italy
   2. Paolo Masi
   3. Italian Wine
   4. Self adhesive label
   5. Chianti

3. 
   1. Bodega Tittarelli - ARGENTINA
   2. The Forefather
   3. Argentinian Wine
   4. Self adhesive label
   5. Limited Edition Gran Reserva. Designed to celebrate the history of the family Bodega.

4. 
   1. Bodega Tittarelli - ARGENTINA
   2. Don Enrico
   3. Argentinian Wine
   4. Self adhesive label
   5. Selección Especial

5.
1. Direct Wines - United Kingdom
2. Casa del Rio Verde
3. Argentinian Wine
4. Self adhesive label
5. Sauvignon Blanc

6.
1. Bodega Tittarelli
2. Tittarelli
3. Argentinian Wine
4. Self adhesive label
5. A special Edition Trophy Reserve Oak Aged Malbec

7. 1. Bodegas Larchago - Spain
   2. Val de Oron
   3. Spanish Wine
   4. Self adhesive label
   5. A modern design to reflect the new Ribera del Duero region of Spain

8. 1. Union de Cosecheros de Labastida
   2. Castillo Labastida
   3. Spanish Wine - Rioja
   4. Self adhesive label
   5. A special edition wine to celebrate the 40th Anniversary of the Bodega.

# Scotch Whisky

9. 
1. UDV - United Kingdom
2. Cragganmore Whisky
3. Scotch Whisky
4. Self adhesive label, box carton
5. One of a range og premium scotch whiskies specifically designed for the duty free market.

10. 
1. UDV - United Kingdom
2. The Glenlivet
3. Scotch Whisky
4. Paper, wooden box, brass.
5. A Special 1969 Vintage for the duty free market.

11. 
1. UDV - United Kingdom
2. Dalwhinnie
3. Scotch Whisky
4. Self adhesive label, box carton
5. One of a range og premium scotch whiskies specifically designed for the duty free market.

# ALOOF DESIGN
## BRITAIN

## aloof
### branding/design consultancy

Aloof are a branding and design consultancy, working on national and international projects across a range of business sectors.

We specialise in graphic and structural design for brand identity, promotion and packaging.

We encourage a creative partnership with our clients, committing to define and communicate brand and product values through efficient and distinctive design

Established in 1999, Aloof remains a limited, independent company owned by our partners, who are both designers. Our studio is based near Brighton, East Sussex.

  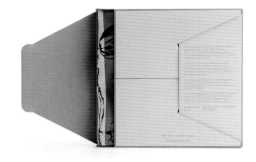

1  Twentytwentyone | UK
   Lucienne Day Teatowel
   Packaging & Graphic Design

2  Twentytwentyone | UK
Robin Day Tray
Packaging & Graphic Design

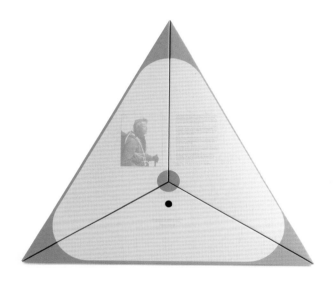

3   Jo Wood Organics | UK
    Everyday Range
    Packaging & Graphic Design

4   U'Luvka Vodka | UK
    U'Luvka Vodka
    Packaging & Graphic Design

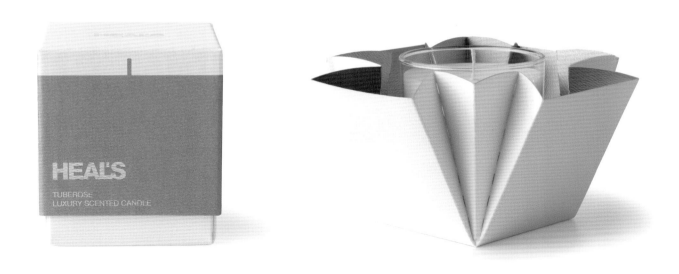

5   Heals | UK
    Luxury Candle Range
    Packaging & Graphic Design

# AUSTON DESIGN GROUP
## USA

## adg | auston design group

Auston Design Group offers full-service package design and brand development capabilities with extensive experience in specialty foods, luxury goods and the wine industry. At ADG, each design assignment is an exercise in branding: creation, enhancement, application and leverage. Our range of branding solutions demonstrates the flexibility and depth of strategic thought needed to create distinctive brand personalities.

As creative director, Tony Auston's approach begins with the simple belief that there is a unique story behind every brand. It is the studio's challenge and commitment to find that story and present it in the most appropriate and affective manner. Mr. Auston's belief is that the process must begin in the real world with analysis of competitive set, market trends and consumer needs; and that successful design happens only when fact is the foundation for creativity. From facts flow informed, visionary design solutions to real-world marketing challenges.

Auston Design Group is located in California's San Francisco Bay Area where the weather is temperate, the food is amazing and the wine flows freely. If you can't visit them at the studio, please visit the website.

austondesign.com
info@austondesign.com

Photography by David Bishop: dbsf@mac.com

1    Five Vintners
      Zinfandel

2    Diageo Chateau & Estates
      Beauzeaux, Red Wine
      Product Launch Kit

3    Diageo Chateau & Estates
      Beauzeaux, Red Wine

4

5

4   Darioush Estate Vineyards
    Duel, Red Wine

5   Joseph Phelps Vineyards
    Ovation, Chardonnay

6   Jim Neal Wine Company
    Chariot, Sangiovese

7   Bogle Vineyards
    Cabernet Sauvignon

8   Raymond Vineyard & Cellar
    R. Collection,
    Cabernet Sauvignon
    Photography: Paul Kirchner

9   Seghesio Family Vineyards
    Italian Varietal Wines

10   Dry Creek Vineyard
     Anniversary Cuvée
     Zinfandel

11   Delicato Family Vineyards
     Gnarly Head
     Old Vine Zinfandel

12   Purple Wine Company
     Rock Rabbit
     Sauvignon Blanc

13  Chinablue
    Chinese Sauces & Marinades

14  Seeds of Change
    Organic Food Line

15  Vitasoy USA
    Azumaya Tofu

16  Fusion Foods
    Napa Valley Verjus

17  Delicato Family Vineyards
    Bota Box, Californian Wines

15

16

17

18 Serendipity Chocolates
   Chocolate Bars

19 Woodhouse Chocolate
   Fine Chocolates
   Photography: Dan Mills

20 Dry Creek Vineyard
   The Mariner, Red Wine

21 Benziger Family Winery
   De Coelo, Pinot Noir

# JOÃO MACHADO
## PORTUGAL

João Machado was born in Coimbra, 1942. Graduated in Sculpture by the Oporto Fine School of Arts.

Individual Exhibitions

- 1986 Art Poster Gallery, Lambsheim, Germany.
- 1987 Annecy/Bonlieu - Centre d'Action France.
- 1989 Lincoln Center, Colorado, EUA.
- 1996 Galeria de la Casa del Poe ta, Mexico.
- 1997 DDD Galery, Osaka, Japan.
- 1998 Casa Garden, Macau.
- 2001 Pécsi Galéria, Pécs, Hungary
- 2002 Dansk Plakatmuseum, Arhus., Danemark.
- 2006 Ginza Graphic Gallery, 20th Anniversary GGG / DDD Project, Japan.
- 2007 International Triennial of Stage Poster, Bulgary.

Awards

- 1989 Special Award "Die Erste Internationale Litfass Kunst Biennale", Germany Bronze Medal "Bienal do Livro de Leipzig", Germany
- 1996 Award, "Computer Art Bienal", Rzeszow, Poland
- 1997 1st Award "Mikulás Galanda, Bienal do Livro de Martin", Slovak Republic; 1st Award "First International Competition for Fair Poster", Bulgary
- 1997 1st Award "Logo Film Commission, Association of film Commissioners International Denver", USA
- 1999 1st Award "Best of Show, European Design Annual", Great Britain; Award Zgraf 8 Icograda Excellence, Croatia
- 2004 2nd Award, "4th International Triennal of Stage Poster", Bulgary.
- 2005 Award "Aziago International Award 2005", for the best worldwide stamp in the Tourism category, Italy
- 2007 Award "Aziago International Award 2007", for the best worldwide stamp in the Protection of the environment category, Italy
- 2008 Honourable Mention EKOPLAGÁT'08, for the collection of four posters

João Machado
Rua Padre Xavier Coutinho, 125
4150-751 Porto Portugal
telf. 351226103772
geral@joaomachado.com

Manuel Alcino, Silversmiths, Porto, Portugal
Manuel Alcino, Tradition and Modernity in Portuguese Silversmiths' Art
Packing-case for book and silver tray
Cardboard and tissue
Graphic design

Ach. Brito , Fajozes, Portugal
Paper bag and soap packing-case
Cardboard
Graphic design

Bacalhoa, Vinhos de Portugal, SA., Azeitão, Portugal
Moscatel de Setúbal Superior 20 Anos
Packing-case
Cardboard
Graphic design

# PROAD IDENTITY
TAIWAN

## PROAD IDENTITY

The essence of design comes from a deep understanding of history.
We see sincerity and depth in design when creativity and innovation are rooted in our own history.

In order to provide brand design strategies at the international level and design standards, PROAD IDENTITY has been working with design teams from international organizations that exchange more international brand knowledge.

Our mission is to optimize the value on every product and every brand by value added from design.

**Proad Identity**
3 Floor, No.42, Section. 2,
Zhongcheng Road,
Taipei 11147, TAIWAN
TEL. +886 2 2833 1943
FAX. +886 2 2835 2948
Mobile. +886 920 290 352
mindy@proadidentity.com

www.proaddesign.com.tw

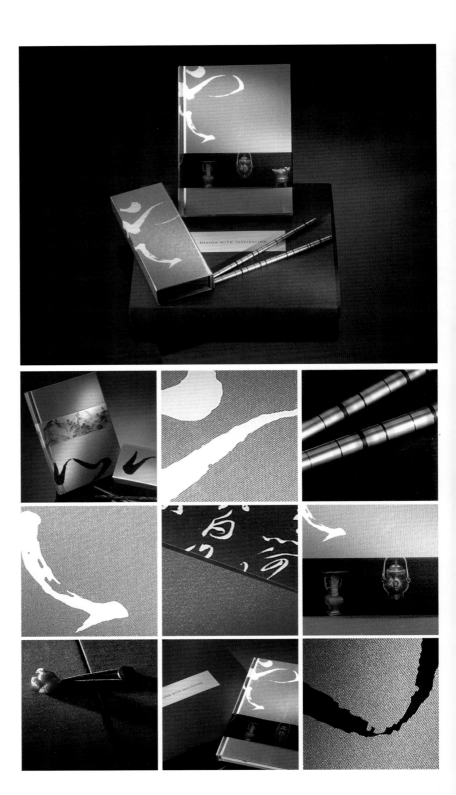

Clint : Proad Identity
Project Name : Stationary Case
Project : Packaging
Award : reddot Design award Winner 2007
      iF commmication daeign Award 2007

# poodehii

Puppet Shows have a deeply instilled style and form-

Culture, craft, toy, and the alluring drama are the elements driving the renewal of the brand consciousness To keep the Poodehii designers in between "traditions" and "personal styles",

The puppet master's hand gesture becomes the symbol and the colors, lines, symbols, concepts, and meta-messages in the facial paintings and costumes are reassembled into new ensembles.

An ingenious idea of the dynamic spiral is integrated into the designs of the book covers to symbolize the front and back stage.

Such versitle mechanism functions as a packaging, stationery, and boutique living-ware and extends into a nostalgic yet modern style of the orient.

Clint : Cheerful Group
Project Name : Poodehii
Project : Brand Identity & Packaging

**Fashionable court tea series
EMPEROR LOVE**

Spirits fly in the tall mountain and put magic on the bamboo leaves. In such a scene of elegance together with the simplified and solemn form and decorations-like the strand as fine as silk or paper as white as sand, like a brocade bag for the imperial jade seal or elegant and pure satin Bestowing a reserved, refined nobility which quietly assuages the public.

The tall mountain representing Oolong speaks of the mellow tea from the tip of mountaintop and corresponds to its brand name Taishan.

The bamboo leaf signifying Pu'er narrates how the aroma and freshness is nourished over time in the air as bamboo leaves dances and breathes with the rhythmic

The floral tea in a package is like the women's rouge box brings a strong nostalgic sense for people to immerse in the belle époque while sipping the tasteful floral tea.

Clint : C-TAI ENTERPRISE CO., LTD.
Product : Tea
Project Name : EMPEROR LOVE
Project :  Packaging
Award :  iF Packaging Award 2008

Clint : C-TAI ENTERPRISE CO., LTD.
Product : Tea
Project Name : EMPEROR LOVE
Project : Packaging
Award : iF Packaging Award 2008

Clint : ONE SIENG TRADING CO., LTD
Product : Cake & Jelly
Project Name : Camellia Banquet Gift Box
Project : Packaging
Award : iF Packaging Award 2007

Clint : Royal Chef International Crop.
Product : Cake
Project Name : Pineapple Cake Gift Box
Project : Packaging
Award : reddot Design Award 2007
         iF Packaging Award 2008

# DESIGNAFFAIRS
## GERMANY

Managing Directors of designaffairs
Nico Michler, Gerd Helmreich, Michael Lanz, Claude Toussaint

designaffairs is a German-based, full-service design agency with studios in Munich and Erlangen. Together with a global network of material, engineering and research partners the team of 44 designers offers a wide range of international design services for well-known brands like Wella, LG, Siemens, Rubbermaid, Haier, WMF and many more.

Besides packaging design, the company offers integrated branding, design strategy, industrial design, user interface design and color & material design.

designaffairs combines its knowledge of clients' brands, various markets, consumer requirements and diverse industrial sectors with design competencies; an interdisciplinary approach that yields sustainable, innovative design solutions. Depending on the project, the design team is formed individually and may consist of, for example, strategists, designers, psychologists, anthropologists usability experts and engineers.

designaffairs is one of the largest design agencies in Europe. In European design award rankings, the agency consistently holds one of the top positions.

The examples on these pages show the packaging design for the innovative Pfeiffer dispenser system.

The Pfeiffer packaging design is a good example of how designaffairs combines the challenge to underline the technical innovation of a product, as well as to convey a high-value and unique appearance.

The Pfeiffer system is an innovative dispenser system, that enables a complete closure of the tube. This innovative closure technique prevents remaining leftovers. Additionally, no bacteria can enter the interior. This high-tech approach is transformed by designaffairs into an elegant, high-class packaging design providing values as genuineness but as well antisepsis via an emotional design. Due to its minimalistic shape the packaging can be adapted easily to different cosmetic brands and their appearance.

1

2

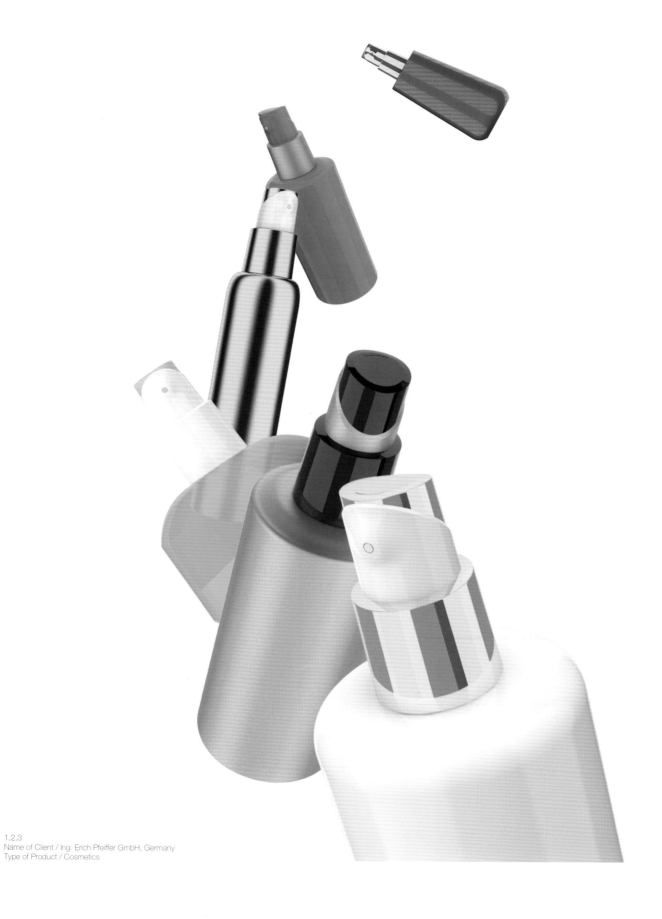

1,2,3
Name of Client / Ing. Erich Pfeiffer GmbH, Germany
Type of Product / Cosmetics

# CHEN DESIGN ASSOCIATES
USA

Chen Design Associates is an internationally recognized, award-winning branding, communications and design firm. For almost 20 years, we've been helping clients find a voice to express the genuine and the necessary, the illuminating and the unexpected. Telling your best story gets your business noticed. Telling it in bold and authentic terms keeps your audience riveted and loyal.

Our work is guided by a strategic and ambitious creative process and is grounded in the particular objectives of each client. We infuse a sense of alignment and engagement into every aspect of our work. We are streamlined to stay nimble and personal, even as our projects increase in size and scope. We intergrate our principles of sustainable design into all our practices and production. Sustainability has always been, and is increasingly more so every day, at the core of our business.

Because we are experts, not egoists, we engage clients in the process to foster respect and fuel invention. Together, we move people — to explore, to wonder, to succeed.

**Chen Design Associates**
649 Front Street Third Floor
San Francisco, CA 94111
United States
T 415.896.5338
info@chendesign.com
www.chendesign.com

**Key personnel** *(from top, left to right)*
Joshua C. Chen, principal & creative director // Kathrin Blatter, senior designer // Margaret Hartwell, brand strategy director // Laurie Carrigan, design director // Max Spector, art director & senior designer // Ryan McAdam, interactive design director // Jeff Plank, business development director // Louise Rice, production manager // Kathryn Hoffman, art director & PR

1 The North Face
Footwear
Packaging design

2  The North Face
   Footwear and Accessories
   Packaging and information design in four languages
   Printed on 100% post-consumer recycled content

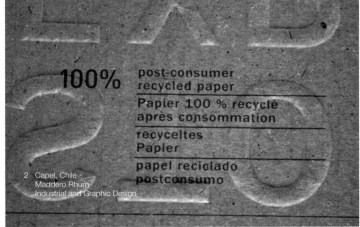

2  Capel, Chile
   Maddero Rhum
   Industrial and Graphic Design

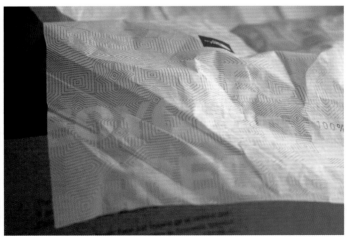

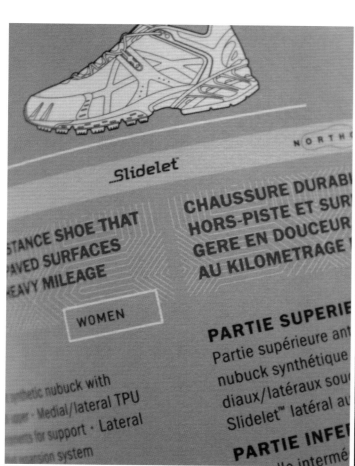

6 Capel, Chile
Maddero Rhum
Graphic Design

3   The North Face
    Global merchandise hang tag system
    Packaging and information design in four languages
    Printed on 100% post-consumer recycled content

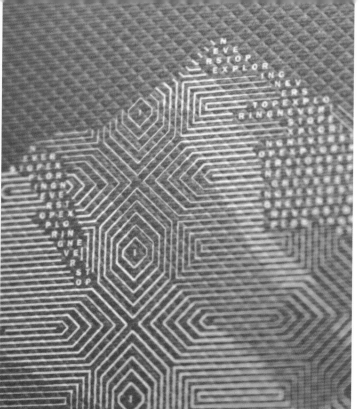

4  *Here on Out*
   Cordis
   CD packaging design

5  *Transcriptions*
   King Cole Trio
   EMI Capitol Records / Blue Note Records
   CD packaging design
   3-CD box set with commemorative booklet

# UP CREATIVE
TAIWAN

Founded in 1988 and keeping constancy in commercial design, UP CREATIVE is neither a commercial agency nor a fine art creation, but a source of professional suggestion and utility creative design, which awarded by lots of international and local competitions. UP CREATIVE firmly believes that commercial design will eventually go back to trade itself, only under the concept of marketing plans and visual design can provide a useful and unique design to reach the most powerful visual effect in greatest benefit. UP CREATIVE always persists in "Creation, Quality, Service, and Timing". Without a qualified and creative design in time equals to the failure in effective service to clients. Her philosophy —progress and creation— are continuous, because she exactly knows what commercial design demands. Following the footstep of new age, she will keep her design in Utmost Progress.

4F. No.269 Sec. 5 Zhongxiao E. Rd. Taipei 11065 Taiwan R.O.C.
**T** +886 2 2765 1181 | **E** up@upcreate.com.tw
www.upcreate.com.tw

1. Herbage International Taiwan Co.
   herbage
   Massage Oil

2. Herbage International Taiwan Co.
   herbage
   Herbal Soap

3. TVJ
   Jplus
   Essential Oil

4. IBL Pharmaceutical Co., Inc.
   IBL
   Shower Gel

5. Billy King Jewellery Co., Ltd.
   BILLYKING
   Skin care series

Fong Nien Fong Ho Enterprises Corp.
Le tea
Fruit Tea   Packaging & label design

Fong Nien Fong Ho Enterprises Corp.
Le Power
Energy Drink

Uni-president Enterprises Corp.
La gauche de La Seine
Cheese Cake

Jing Mai Lang Food Corp.
Jing Mai Lang
Instant Noodles

|1| |3|4|5|
|---|---|---|---|---|
|2| |6| | |

1. Jian Jan Tea Co., Ltd.
   Tian Chih Tea
   Loose Tea

2. Jian Jan Tea Co., Ltd.
   Le Midi Hotel Tea
   Loose Tea

3. Chou Chin Industrial Co., Ltd.
   Natural Nuts
   Biscuit

4. Hunya Foods Co., Ltd.
   Wedding Cake series

5. Hunya Foods Co., Ltd.
   Wedding Cake series

6. Hunya Foods Co., Ltd.
   Xi Yue
   Moon Cake

# MARTA ROURICH
## SPAIN

Diseño Gráfico
MARTA ROURICH

Diseño Gráfico
MARTA ROURICH
Putxet 80-82 1º A • 08023 Barcelona
Tel. +034 93 212 31 33  Fax +34 93 418 81 23
mrourich@snap3.com
SPAIN

Founded in 1982. The philosophy of Rourich studio is "design to communicate" mantaining a high standard of quality for its selected clients.

A highly identified team of specialist associates and a direct relationship with their clients, in such a way that the person who receives the assignment is the same one who thinks up, designs and presents the work, one of the main characteristics of this studio.

Rourich works mainly in packaging design, sales displays, catalogues and corporative image.

Some of its clients are:
CATAGourmet, COASA, Quely S.A., Sheraton Algarve, Acdha, Champagne Brice, Leciñena, Cosa, International Financial Advisors, Rellman Food, UVIPE, Aguas de Fuensanta, G.I. Architects and Designers, Gremio de Panaderos de Barcelona, Fundación Ared, Quesera del Cares.

Vino Rioja/ Rioja Wine (D.O. Rioja)

Platos cocinados/Cooked dishes (CataGourmet)

Paté de mariscos/ Seafood spread (CataGourmet)

Patés cárnicos/Meat spread (CataGourmet)

Mermelada/Jam (CataGourmet)

Miel/Honey (CataGourmet)

Queso artesano/Artisan cheese (CataGourmet)

Embutidos de ciervo/Cold venison meat (CataGourmet)

Embutidos de jabalí/Cold wild boar meat (CataGourmet)

Aceite, Vinagre y Setas/Olive oil, Vinager and Wild mushrooms (Cata Gourmet)

# DEZIRO
JAPAN

Deziro has been continuously involved in consumer brand identities and packaging development since 1990. Keeping the idea of brand identity as our base, we aim to create packaging that accurately expresses the unique traits of products to consumers.

Our design concept is to develop packaging that builds brands while also selling products.

**DEZIRO Co., Ltd.**
5-14-19, Roppongi, Minato-ku, Tokyo, 106-0032
Tel: +81-3-5572-0771   Fax: +81-3-5572-0772
http://www.deziro.co.jp
E-mail: deziro@deziro.co.jp

1 Nisshin OilliO Group
"Shizuku"
"Canola Oil"
Graphic Design

2  Japan Food & Liquor Alliance Inc. Food Sales
Morita "Blue-ribbon" series
Sauces
Graphic Design

3 FANCL
The germinated brown rice gruel
Gruel
Graphic Design

4  Nippon Milk Community
  "Genuine Fruit" series
  Fruits Juice
  Graphic Design

5 Nippon Milk Communitty
 "Plentiful Milk" yogurt
 Yogurt
 Graphic Design

# blackandgold
FRANCE

From the left to the right are : Daniel Dhondt, President.
Cécilia Tassin, Associate Director / Strategy and New Business.
Yannick Soubrier, Associate Director / Creation.
Mélanie Gransart, Associate Director / Creation.
Laurent Ferragu, Associate Director / Creation.
Elodie Orieux, Associate Director / Marketing.
Nicolas Julhiet, Associate Director / Marketing.

Like any living organism, a brand needs to evolve and adapt to constantly changing environments. With 60 specialists based in Paris and Shanghai, and an integrated range of marketing, research and design services, Balckandgold helps you to anticipate changes and inspire your innovation process.

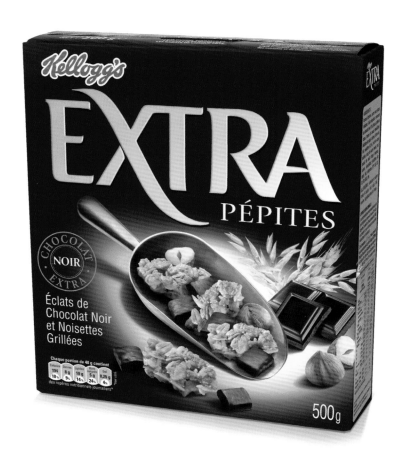

Kellogg's Extra : brand and packaging revamp
Kellogg's Spécial K : packaging upgrade
Badoit : premium bottle creation
Belin Les Petites Recettes : brand and range creation

Joker Vital Protect : brand and packaging design
Agrokor Smart : concept, naming, brand and package design creation
Yoplait Perle de Lait : packaging renewal
Yoplait Petits Filous Tub's : packaging upgrade and mascot creation

Evian Brumisateur : Fashion Limited Edition Creation
Sonatina : Retail Concept Creation and Implementation
Yves Rocher Sérum Végétal : Brand, packaging and structural redesign
Dim : men's range revitalization

# LLOYD GREY DESIGN
## AUSTRALIA

### BACKGROUND
Based in Brisbane, Australia, Lloyd Grey Design is a visual communication consultancy that stretches beyond the confines of ordinary expectation.

We craft finely-honed strategic design responses to diverse communication challenges. Our scope includes branding, corporate identity, packaging, information design, annual reports and web creation.

In fifteen years, our loyal clients have ranged from large corporations to small businesses. Our approach transcends industry barriers, innovating for every sector.

### OUR PHILOSOPHY
Our philosophy can be encapsulated in two words. Connect. Resonate. We connect with our clients' worlds. We listen to their business goals. We walk a mile in their customers' shoes. The result is creative partnership driving intelligent solutions.

We combine research, intuition and inspiration to ensure every design resonates with its target market.

### OUR APPROACH TO PACKAGING
Packaging is the pivotal moment of consumer marketing. It challenges designers to achieve more with less.

We search for the perfect balance of images, colours, moods and emotive associations to create the vibrant point where brand characteristics and customers' aspirations meet.

This fusion allows unique brand impressions to emerge and imbues the product with values that inspire buying decisions.

**www.lgd.com.au**

1  Bunnyconnellen, Australia
   Bunnyconnellen
   Gourmet olive oil range
   Branding, label design

  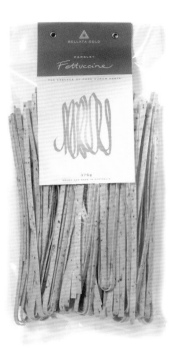

2   Backwell Trading Company, Australia
    Bellata Gold
    Gourmet pasta range
    Branding, label design

3   Dennar Pty Ltd, Australia
    Good Morning Cereals
    Organic breakfast cereal range
    Branding, packaging design

4   Bunnyconnellen, Australia
    Bunnyconnellen
    Gourmet table olives range
    Branding, label design

5   Mother Meg's Fine Foods, Australia
    Mother Meg's festive range
    Branding, structural packaging
    design and label design

# DOLHEM DESIGN
## SWEDEN

Dolhem Design's head office is located in the heart of Stockholm

Dolhem Design, create, refine and clarify our clients' brands, products and services through a mixture of quality, creativity, charisma and economical thinking (in the order preferred by the client). Even if we use different models depending on client need and objectives, design management is always the common denominator.

We have built up a broad client base that spans a large range of different businesses in both the public and private sectors since 1998. As a result of belonging to INAREA, an international design network, we have access to expertise in different markets and an inexhaustible source of information that can be used on behalf of our clients. Dolhem Design's international character can also be seen in the fact that five languages are fluently spoken at the company (Swedish, English, French, and Czech).

Feel free to order our case stories or to book a meeting with us so that we can explain in detail how we work and look upon branding as well as how design management can be seen as an investment rather than only as an expense.

Please visit: www.dolhemdesign.se

Dolhem Design
Nybrogatan 3
SE-11434 Stockholm
Sweden

Tel: +46 (0)8 661 50 47
Fax: +46 (0)8 661 50 48

info@dolhemdesign.se
www.dolhemdesign.se

**Estelle & Thild Ecorganic baby & child** provides the first Ecocert certified range of baby care products to the Swedish market. A playful yet clean Scandinavian manner was applied on the packaging to appeal to aware parents with young children. The pattern of avocado leaves reappears throughout the Estelle & Thild profile, not only on the packages, as avocado oil is one of the most frequently used ingredients in the range. The pattern also symbolizes the natural origin of the products. The products are differentiated from eachother by the mild pastel colours, often associated with children's products. Visually the product range stands out among its Swedish competitors.

# NO.PARKING
## ITALY

above, left to right: matteo grotto_elisa dall'angelo_giotto antonio andolfatto martinez_
caterina romio_sabine lercher_anna saraconi_caroline mohr

a magnificent art nouveau staircase in a seicento palace. this is our studio. there's no parking space. design objects, a lot of papers, magazines, books and an antique ceiling fresco. if you like this maybe it means that you even like us: anna, caroline, caterina, elisa, giotto, matteo, sabine. if you come to see us leave your car outside the town and catch the bus.

we love graphics, giving the visual communication a concept, designing logos, websites, books and packaging, its all about communicating and making the world more beautiful.

**www.noparking.it**

**no.parking**
Contrà S. Barbara, 19
36100 Vicenza
Italy
T/F+39 0444 327861
info@noparking.it
www.noparking.it

1 Cartindustria Veneta | Italy
  Moon · compact toilet paper
  > design of a portable toilet paper holder
  > design of a starter kit including 6 rolls and dispenser
  > label design

2  JMT Leather | UK | Italy
packaging for leather collection

# SiO Design
## JAPAN

SiO DESIGN

Designing Smartly.
This is the core ideal of SiO DESIGN.
Amidst the turbulent changes of these times, to offer designs that cut away all waste to provide the ideal posture for the future, that is what we see as our mission.

In package design, the issue is how to visualize the product concept and convey it to the consumer. What is required is the capacity to hold the maker's IDENTITY and to communicate the features of the product effectively. Too much expression (information) would blur the focus and dull the force of communication. Based on visual design of the information quantity to match the scale, we imagine a package that provides that something extra in its image.
That something extra is the "what it resembles„ "the appearance of tasting good„ and other elements entrusted to the sense of the consumer.

http//:www.sio-design.co.jp
info@sio-design.co.jp

CLIENT : MEIJI SEIKA KAISHA, LTD. Japan
PRODUCT NAME : Hot Chocolat

CLIENT : DIAGEO Japan K.K. / PRODUCT NAME : SMIRNOFF Shirakaba no Megumi

CLIENT : UCC UESHIMA COFFEE., LTD., Japan
PRODUCT NAME : Café Bruno

CLIENT : UCC UESHIMA COFFEE., LTD., Japan
PRODUCT NAME : Americano BLACK

CLIENT : Coca-Cola (Japan) Company, Limited
PRODUCT NAME : Schweppes

©The Coca-Cola Company

CLIENT : Coca-Cola (Japan) Company, Limited
PRODUCT NAME : Sokenbicha

©The Coca-Cola Company

# TUCKER CREATIVE
## AUSTRALIA

**Tucker Creative - achieving excellence in packaging for over 30 years.**

Tucker Creative is a leading Australian design, marketing, advertising, branding and packaging specialist.

Our focus is on successful outcomes for our Australian and International clients. The Tucker Creative advantage is our ability to create and design a brand from conception – right through to enjoying resounding market success.

Our unique process and experience of creative strategic thinking and assessment combined with creative design and communication solutions ensure innovative outstanding and inspiring work.

Tucker Creative has accumulated a huge folio of work, prestigious international awards and a proven track record for providing successful and compelling packaging and branding solutions.

Over the next eight pages, we hope you enjoy a cross section of our more recent work. For further examples, or for more information, please call +61 8 8331 1700 or visit www.tuckercreative.com.au

**Head Office & Studios**
57c Kensington Road
Norwood South Australia 5067

**T** +61 8 8331 1700
**F** +61 8 8331 1222

**USA**
**T** +1 707 224 5670
**F** +1 707 224 5683

tuckers@tuckercreative.com.au
**www.tuckercreative.com.au**

1 · Oliver's Taranga Vineyards, Australia
· Expatriate
· Wine
· Paper, glass, tin, cardboard
· Winemaker's travel diary

3 · TeAro Estate, Australia
· TeAro Estate Iron Fist, Joker's Grin and Barefooter
· Wine
· Paper, glass, aluminium

4 · TeAro Estate, Australia
· TeAro Estate Two Charlies and Minnie & Elsa
· Wine
· Paper, glass, aluminium

5 · TeAro Estate, Australia
· TeAro Estate The Pump Jack and The Charred Door
· Wine
· Paper, glass, aluminium

2 · Wirra Wirra Vineyards, Australia
· Wirra Wirra Mrs Wigley
· Wine
· Paper, glass, aluminium, stainless steel

6 · Constellation Wines Australia, Australia
· Leasingham
  Bin 8 K.S. Riesling
· Wine
· Paper, glass, aluminium

7 · Constellation Wines Australia, Australia
· Leasingham Individual Vineyard range
· Wine
· Paper, glass, aluminium

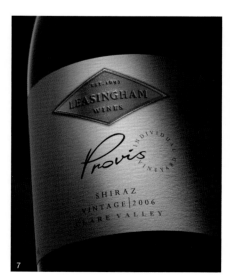

8 · Sevenhill Cellars, Australia
 · Sevenhill Lost Boot
   and White Spider
 · Wine
 · Paper, glass, aluminium

9 · Sevenhill Cellars, Australia
 · Sevenhill Inigo
 · Wine
 · Paper, glass, aluminium

10 · Sevenhill Cellars, Australia
 · Sevenhill St. Aloysius
   and St. Ignatius
 · Wine
 · Paper, glass, aluminium

11 · Sevenhill Cellars, Australia
 · Sevenhill Brother John May S.J.
 · Wine
 · Paper, glass, aluminium, cardboard,
   woven ribbon

12 · The Two Metre Tall Company, Australia
· Derwent Clear Ale, Forester Pale Ale,
  Huon Dark Ale and Cleansing Ale
· Beer
· Paper, glass, steel, cardboard

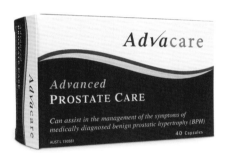

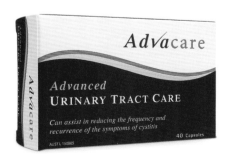

13 · Biologic, Australia
 · Advacare
 · Neutraceutical
 · Cardboard, foil

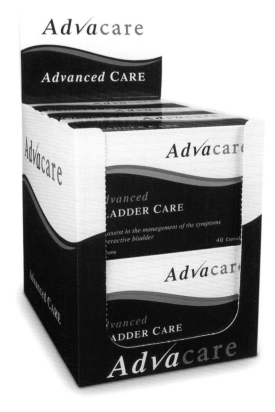

14 · Misha's Vineyard, New Zealand
 · Misha's Vineyard range
 · Wine
 · Paper, glass, aluminium, cardboard

15 · Rodney Strong Wine Estates, USA
 · Rockaway
 · Wine
 · Glass, ceramic print, tin, glass

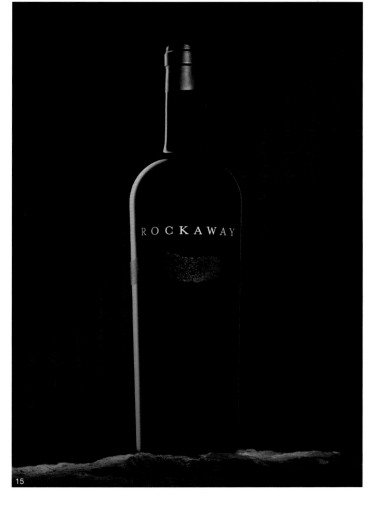

# STUDIO GT&P
## ITALY

# Studio GT&P

Studio GT&P is a multi-disciplinary visual communications firm, founded by Gianluigi Tobanelli in 1985. We work at a human scale, building long-standing partnerships with our clients. We believe that good design helps businesses clarify and realize their vision, enhance their products and services and serve their customers better.

We provide the following services:
- **Identity** (Logo design, Stationery, Signage, Style Manuals)
- **Packaging** (Package and Brand Identity Design)
- **Print** (Annual Reports. Company Profiles, Brochures, Product Catalogues, Newsletters & Periodicals. Direct Mail, Flyers, Promotional Material)
- **Interactive** (Website and Web Collateral Design and Development)

**Studio GT&P**
Via L. Ariosto, 5
06034 Foligno (PG), Italy
Tel. +39 0742 320372
info@tobanelli.it
www.tobanelli.it

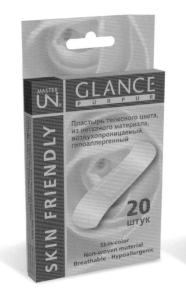

1

1  Pharmline Alliance LLC, UK
   Glance
   Plasters
   Box
   Package design

2  Agricola Spacchetti, Italy
   Colle Ciocco
   Sagrantino Wine
   Glass bottle / Paper labels
   Label design

3  Agricola Spacchetti, Italy
   Colle Ciocco
   Sagrantino Passito Wine
   Glass bottle / Paper labels
   Label design

2

3

4   Agricola Spacchetti, Italy
    Colle Ciooco
    Rosso di Montefalco Wine
    Glass bottle / Paper label
    Label design

5   Agricola Spacchetti, Italy
    Colle Ciooco
    Clarignano Wine
    Glass bottle / Paper label
    Label design

6   Cantina Ricciolini
    Borgovivo
    Sagrantino Wine
    Glass bottle / Paper labels
    Label design

7   Sanpotente
    Sanpotente
    Rosso di Montefalco Wine
    Glass bottle / Paper label
    Label design

# KHDESIGN GMBH

KHDESIGN GMBH, OFFENBACH, GERMANY

*Creating & leading brands*

### ACTIVE BRANDS
Dynamic markets demand strong brands. But these same brands must remain flexible. A lasting presence in the market demands the ability to adapt. Flexible brand evolution strategies enhance product loyalty and capitalise on previous user experience. A successful brand is a living organism, interacting with its environment in myriad ways: this activity is the source of the brand's vitality in the marketplace.

### MOBILE CONSUMERS
Society today is more mobile than ever before. This is true not just in a physical, geographical sense: it applies equally to forms of expression, changing fields of interest and responsibility; to today's dynamic lifestyles. This mobility is reflected in the choices people make, in their identification with products. Trend controlling is one of the methods we integrate in any brand management strategy, in order to predict change and offer pro-active guidance.

### HOLISTIC VISION
The basis for evolutionary brand management is an integrated system, encompassing all the elements which together project the image and qualities of a brand. Core values must remain identical, irrespective of whether the brand is communicating via POS or a complex multimedia presentation. Only a brand architecture that meets these challenges can go on to adapt to changing market requirements and truly add value to a product.

khdesign gmbh
Lilistraße 83 D/09
D-63067 Offenbach
Phone: +49 (0)69/97 08 05-0
Fax: +49 (0)69/7 07 83 71
E-Mail: info@khdesign.de
Internet: www.khdesign.de

MEMBER OF
GLOBAL
DESIGN
SOURCE
EXPERTS IN
BRANDING &
PACKAGING
www.g-d-s.net

1. Schwarze & Schlichte
2. Sweet Lips
3. Fruit Liqueurs
4. Paper
5. Young and trendy fruit liqueur in three different flavours
(Illustrations by Karin Ecker-Spaniol)

1. Terra Mundo
2. Hericium / Reishi / Maitake / Shiitake
3. Vital mushrooms
4. Coated paper board
5. Packaging design for traditional mushroom products

1. dm-drogerie markt
2. Prinzessin Sternenzauber
3. Body Care products for girls
4. Foil on bottles
5. Coltish packaging design to attract young girls

1. dm-drogerie markt
2. alverde decorative Naturkosmetik
3. Natural make up Cosmetic serial
4. Various materials (foil on containers)
5. Unique packaging design with special colour and surface finish

# DESIGN FORCE
### JAPAN

www.group-force.com
info@group-force.com

**"Creativity Changes The BUSINESS."**

DESIGN FORCE aims at providing creative design for corporate identity, branding, product and package, advertisement, and web sites.
Our design brings brand loyalty in the eyes of your customers and promises to incubate all new & big businesses.

KOKUYO S&T / ID NAME card holder series

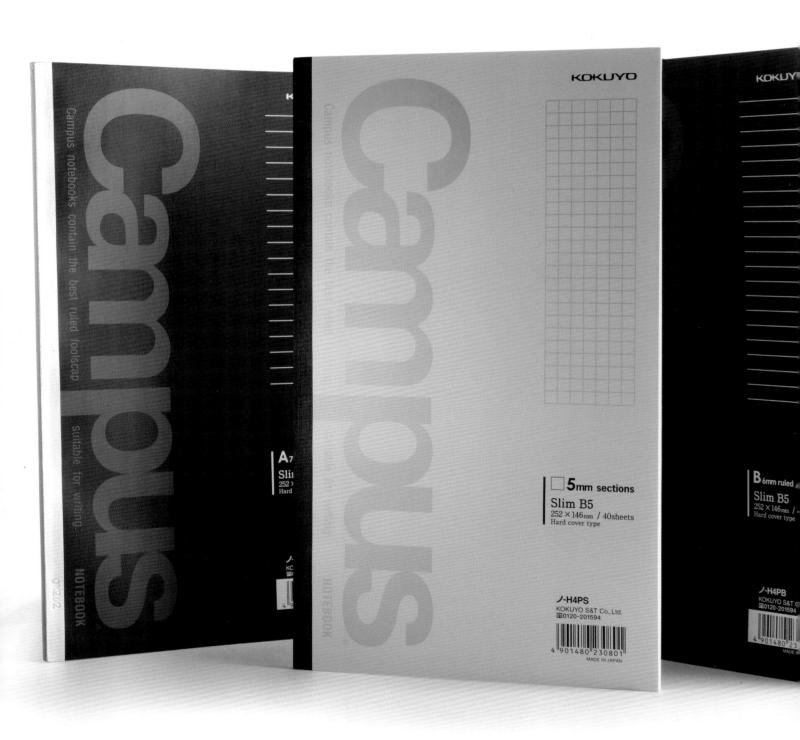

**KOKUYO S&T / "Slim B5"size note**

Aprica Children's Products / Baby's detergent

**PINE / Candy**

**Tominaga Boeki Kaisha / Coffee**

# MILLER CREATIVE LLC
USA

# MILLER

Miller Creative LLC is a branding and packaging design consultancy based in New Jersey. Yael Miller is the creative force behind the consultancy. She has over nine years experience in branding and in packaging design and production, primarily for food products, luxury goods, cosmetics and gourmet confectionary brands.

Miller Creative's international clients span several countries, including Australia, Bahrain and India.

From very small start-up companies to large-scale international ventures, Miller Creative has the flexibility and the expertise to bring its clients' brands to market with strategic insight.

Not only is Miller Creative adept at developing world-class brand identity and packaging visuals for its clients, but this small studio also brings manufacturing know-how to the table. Miller Creative advises its clients on issues relating to production of paper-based packaging, film, thermoformed plastics, sustainable materials and specialized materials used in luxury goods and gourmet confectionary product packaging.

The studio has the flexibility to operate across multiple time zones worldwide - coordinating designers, clients and production vendors in various countries.

The studio's design work continues to win design awards and has also been published internationally in respected design publications.

www.yaelmiller.com

1 - Beautao Active Botanicals packaging designed by Miller Creative LLC

Miller Creative LLC
5 Sharon Court
Lakewood, New Jersey 08701
USA
T +732 600 3933
yael@yaelmiller.com
www.yaelmiller.com

2 - Conflict of Interest Winery
Shirah Syrah Wine Label 2005 Vintage
Concept and label design
Letterpress ink on label

Photo Credit: Andre Jackamets

3 - Beautao Active Botanicals
Skincare packaging collection
Brand naming, identity and packaging design
Custom-colored ink on folding paperboard

Photo Credit: Andre Jackamets

Photo Credit: Andre Jackamets

4 - Crepini Café
Gourmet frozen filled crepes
Redesign of brand identity and packaging design
Printed on unbleached kraft paper

5 - Crepini Café
Gourmet frozen filled crepes
Redesign of brand identity and packaging design
Printed on unbleached kraft paper
Illustration Credit: Roger Xavier

# BRUNAZZI&ASSOCIATI
## ITALY

Above: Giovanni&Andrea Brunazzi, Chairman&Creative Director

Created in 1985 by Giovanni Brunazzi, one of the pioneers of company communication and corporale image in Italy, Brunazzi&Associati is one of a small number of Italian advertising agencies that specialise in corporate identity, publishing and packaging, which are able to provide exclusive, highly professional services in the field of integrated communication and image strategies.

Directly, or with the support of specialist external structures, Brunazzi&Associati coordinates its customer's needs with specific and wide ranging assistance, from industrial and product design to management of integrated communication projects, conventional advertising campaigns, the realisation of new package design solutions for brand product or private labels, the design of stands and promotional items, external relations and graphic projects for books, magazines, posters, quality pubblications and web site.

Every action plan studied by Brunazzi&Associati always depends on detailed examination of the customer company's existing image, accompanied by analysis of its manifacturing and distributional set-up, the competition and its reference market.

Many Brunazzi&Associati projects have received prize and award, not only in Italy.

Brunazzi&Associati
Via Andorno 22
10153 Torino, Italy
tel + 39 011 812 5397
fax +39 011 817 0702
www.brunazzi.com
info@brunazzi.com

Above:
1. Via delle Indie | Italy
2. Poesie Mediterranee
3. Dehydrated sauce
4. Flexible packaging
5. Still-life photos showing the ingredients

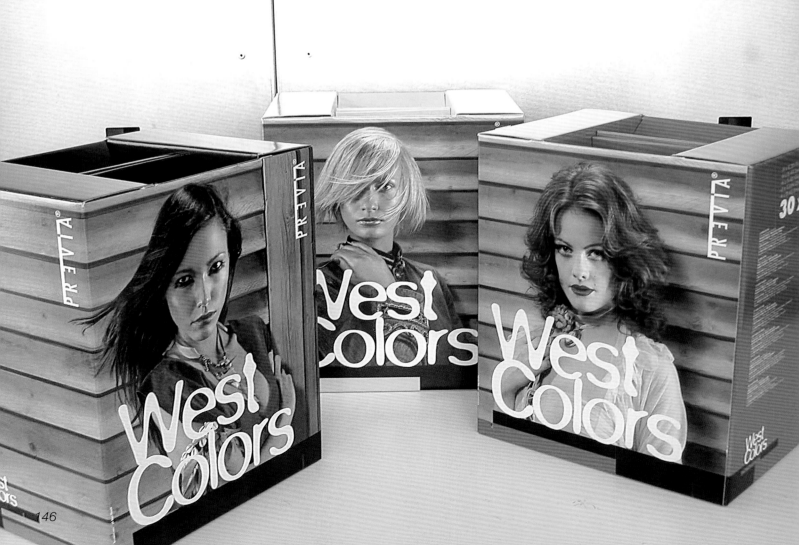

1 Interprogress | Italy
2 Previa
3 Hair cosmetics
4 Plastic, aluminium and cardboard
5 Minimalist design based on typography

Above:
1 Cantine Luzi Donadei Fabiani | Italy
2 Donna Cecilia
3 High quality Wine
4 Label
5 Design based on calligraphy

Below
1 Canonica di Corteranzo | Italy
2 Isabella
3 High quality Wines
4 Labels
5 Minimalist design based on typography

Right:
1 Terre da Vino | Italy
2 Malvasia / Asti
3 High quality Wines
4 Labels
5 Graphic design
  based on classic typography

# LITTLE BIG BRANDS
## USA

**Little Big Brands**
38 High Ave., 4th Floor
Nyack, NY 10960 USA
845.480.5911
pam@littlebigbrands.com
www.littlebigbrands.com

Little Big Brands has been creating exceptional brand and package design in the metropolitan New York City area since 2001. Our work is insightful, inspired, and never frivolous, and we take great pride in finding the nuances that make our clients unique. We believe that great design begins with fearless clients who have the desire to do what's truly right for their brand. Be brave, give us a call.

Little Big...for big brands that want to be bigger and little brands that aspire to be big.™

1   Lornamead, United States
    Finesse
    hair care products
    pressure sensitive labels, steel and aluminum cans
    identity & package design

Lornamead, United States   2
Yardley
natural bar soap
paperboard
identity & package design

151

The Lion Brewery, United States 3
Lionshead
beer
wet-strength paper, steel, paperboard, corrugated cardboard
identity, structure & package design

4   The Lion Brewery, United States
    Stegmaier
    beer
    wet-strength metalized paper, steel, paperboard, corrugated cardboard
    identity, structure & package design

5th & McGraw, United States   5
5th & McGraw
wine
paper, foil
identity, structure & packaging

# DIL BRANDS
## BRASIL ARGENTINA CHILE MEXICO

CORPORATE & CONSUMER BRANDING

# DIL BRANDS

Founded in 1961, DIL Brands is dedicated to strategic branding and design for consumer and corporate brands.

We are able to manage entire branding programs, from structured brainstorming sessions, ideation and concept sketching, to product concept (product, graphic and industrial design) and precise, coordinated implementation.

If there is a need for focused and structured creativity, locally of globally, DIL Brands has all the tools and expertise to do it.

- More than 5,000 new packages over 45 years
- 4 live-connected offices: Brasil – Chile – Argentina – México.
- Multicultural, multiracial, multimotivated
- In-house prototyping capabilities
- Videoconference capabilities
- The most awarded consultancy in our region - Worldstars, Clios, Pentawards, London Awards;
- A CBX partner

Rather than dedicating ourselves to creating simply beautiful and attractive designs, we are focused on delivering "designs that sell". This is the intriguing nature of our business and the essence of our quality standards.

CBX (www.cbx.com) is our worldwide resource, a network of world-class brand strategy and design consultancies that stretch our reach to Europe (UK and The Nederlands), United States (New York, Minneapolis and San Francisco), Oceania (Melbourne) and Asia (Ghengzhou and Shangai – China). Whatever the need, we will speak the language.

SAO Al. Rio Negro 1030 - cj. 2304 - 06454-000 Alphaville SP, Brasil (55 11 4191 9711)
MEX Jaime Balmes 11 Piso 7 Polanco Chapultepec, México (52 55 5580 4047)
SCL Andrés Bello 2777 – of. 2403 Las Condes Santiago, Chile (56 2 594 7878)
BUE Sinclair 2949 Piso 8 C 1425BXO Buenos Aires, Argentina (54 11 4139 5485)
www.dilbrands.com

1. FEMSA Brasil
2. Summer Draft Beer
3. Industrial and Graphic Design

1. Sigma Alimentos - Mexico
2. YOP Drinkable Yoghurt
3. Graphic Design

1. Colgate Palmolive - USA
2. Suavitel Momentos Mágicos - Fabric Softener
3. Industrial and Graphic Design

1. Unilever Foods - Chile
2. Club Ceylán Teas
3. Graphic Design

1. Unilever Foods, LATAM
2. Hellmann's Ketchup
3. Industrial and Graphic Design

1. Jeronimo Martins - Portugal
2. Pingo Doce - Nectars
3. Graphic Design

1. Jeronimo Martins - Portugal
2. Pingo Doce - Frozen Pizzas
3. Graphic Design

1. Jeronimo Martins - Portugal
2. Pingo Doce - Desserts
3. Graphic Design

1. Jeronimo Martins - Portugal
2. Pingo Doce - Nectars
3. Graphic Design

1. Watt's - Chile
2. Watt's Soy
3. Graphic Design

1. Nestlé - Chile
2. Sahne Nuss Chocolate
3. Graphic Design

1. Nestlé - Chile
2. Sahne Nuss Gift Box
3. Industrial and Graphic Design

1. Bimbo - Chile
2. Ideal White and Wholegrain Bread
3. Graphic Design

1. Bimbo - Mexico
2. Plus Vita Cereal Bar
3. Graphic Design

1. Bimbon - Argentina
2. Fruits Cookies Bar
3. Graphic Design

1. Bimbo - Brasil
2. Plus Vita - Integral Snack
3. Graphic Design

1. Coca Cola - Argentina
2. Cepita Functional Products
3. Graphic Design

1. Nestlé - Chile
2. Museo Cookies
3. Graphic Design

1. Nestlé - Chile
2. Trencito Powder Milk Bar
3. Graphic Design

1. Nestlé - Argentina
2. Svelty - Low Cal Ice Cream
3. Graphic Design

1. Nestlé - Argentina
2. Heaven Unik - Ice Cream
3. Graphic Design

1. Nestlé - Chile
2. La Cremería - Premium Ice Cream
3. Graphic Design

# TRIDIMAGE
ARGENTINA

# tridimage
## 3D BRAND & PACKAGE DESIGN

Tridimage is an integrated graphic and structural packaging design consultancy, based in Buenos Aires, Argentina. We have been producing creative and distinctive branding solutions for clients worldwide since 1995.

We possess a blend of strategic thinking, boundless 3D creativity, commercial savvy, and flawless execution. Our extensive experience in 2D and 3D packaging spans across all industries, including food and beverage, home care, health and beauty, technology and durable goods.

Our packaging designs are backed up by our sound knowledge of manufacturing processes, materials and cost constraints, giving our clients a point of difference in an evermore competitive marketplace. At Tridimage we believe that successful packaging reflects the brand positioning, stands out, has conviction and is cost effective in its application.

We run all of our design projects, many of them for clients around the world, from our Buenos Aires studio. Today's technology allows the communication of ideas, creative work and instant access to local market intelligence necessary for the smooth running of a design project from global to local.

Echeverría 3856
C1430BTL - Buenos Aires
Argentina

TEL +54 11 4554 1812
FAX +54 11 4554 3208
SKYPE tridimage

info@tridimage.com
**www.tridimage.com**

1   Grupo Berro / Danone Waters | Spain
    Lanjarón Mineral Water
    PET bottle / OPP label
    Structural Design by Tridimage
    Graphic Design by Grupo Berro
    **SILVER PENTAWARD 2008**

2  Bodega Ikal | Argentina
Ikal 1150 Wines
Brand Identity & Graphic Design

3 Bodegas Etchart | Argentina
Cafayate Wines
Structural & Graphic Design

4 Bodega El Esteco | Argentina
Elementos Wines
Structural & Graphic Design

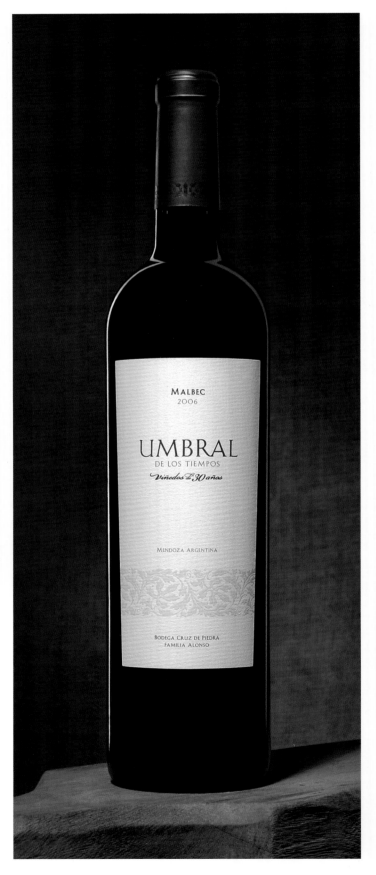

5   Bodega Cruz de Piedra | Argentina
    Cruz de Piedra Blend
    Brand Identity & Graphic Design

6  Bodega Cruz de Piedra | Argentina
Umbral de los Tiempos Wines
Brand Identity & Graphic Design

7  Grupo Berro / Grupo Osborne | Spain
Alma de Magno Brandy
Structural Design by Tridimage
Graphic Design by Grupo Berro

167

8  Laboratorios Firenze | Argentina
   Lizhara Cosmetics
   Brand Identity & Graphic Design

9  Embotelladora Cactus | Mexico
   Cactus Cross Isotonic Beverages
   Structural & Graphic Design

10 Nestle Waters | Argentina
   Eco de los Andes Mineral Water
   Structural & Graphic Design

11 Tau Delta | Argentina
  Tau Delta Dehydrated Foods
  Brand Identity & Graphic Design

12 Embotelladora Cactus | Mexico
  Cactus Wok Iced Green Teas
  Structural & Graphic Design

# BRITTON DESIGN
USA

1  Clos du Val Winery, USA
   Wine Packaging

**P**atti Britton designs for wine and lifestyle clients around the world. She expresses the essence and spirit of each brand with the kind of look and feel that captures the imagination of consumers in a crowded marketplace.

**B**ritton Design is headquartered in picturesque Sonoma, California. Since 1990, Patti Britton has attracted clients not only throughout her home state, but from Italy, Chile, and Australia, as well. Her work for Robert Mondavi, Opus One, Galante, Viansa, Clos du Val, DFV Wines (Delicato), Banfi Vintners, Ferrari-Carano, Remy Amerique/Antinori has won numerous international and national awards, including trophies from the Brand Design Association in New York. For Viansa alone, she won 86 design awards, including the 1997 London International Advertising Awards for Viansa's Athena, voted *The Best Wine Packaging Of The World*.

**P**atti Britton's designs have been featured in books published by *Graphis Bottle Design* and the American Institute of Graphic Arts, plus such magazines as *Critique* and *Communication Arts*.

BRITTON DESIGN
724 FIRST STREET WEST
PO BOX 1653
SONOMA, CA 95476
TEL 707.938.8378
email: pb@brittondesign.com
www.brittondesign.com

**DISTINCTIVE WINE PACKAGING DESIGN THAT INFORMS, ENHANCES AND EXCITES**

2  Opus One Winery USA
   Brochure Series of 3 ~
   Family, Winemaking, and
   Architecture & Design

3  Roy Estate, USA
   Wine Packaging

4  Hermitage Road, Australia
   Wine Packaging

5  Kenneth Volk Vineyards, USA
   Wine Packaging

**DISTINCTIVE WINE PACKAGING DESIGN THAT INFORMS, ENHANCES AND EXCITES**

6 Clos Pegase Winery, USA
  Wine Packaging

7 Stryker Sonoma Winery, USA
  Wine Packaging

8 Galante Vineyards, USA
  Wine Packaging

9 Antinori Wines, Italy
  Estates wine brochure

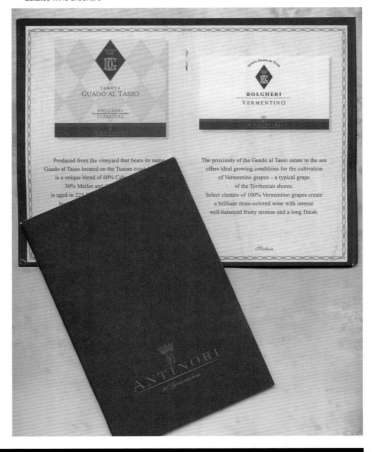

**DISTINCTIVE WINE PACKAGING DESIGN THAT INFORMS, ENHANCES AND EXCITES**

10   Robert Mondavi Winery, USA
     Wine Packaging

11   Opus One Overture, USA
     Wine Packaging

12   Sena, USA and Chile
     Wine Packaging

13   Opus One Winery, USA
     Limited edition fine art print

**DISTINCTIVE WINE PACKAGING DESIGN THAT INFORMS, ENHANCES AND EXCITES**

14  DFV Wines, USA
   Wine Packaging

15  Plata Wine Partners, USA
   Wine Packaging

17  Plata Wine Partners, USA
   Wine Packaging

16  Plata Wine Partners, USA
   Wine Packaging

Designs this page~
Naming, Branding, Design
for the wine industry

18  Plata Wine Partners, USA
   Wine Packaging

**DISTINCTIVE WINE PACKAGING DESIGN THAT INFORMS, ENHANCES AND EXCITES**

19  Viansa Winery, USA
    Athena
    Wine Packaging

20  Viansa Winery, USA
    Frescolina
    Wine Packaging

21  Viansa Winery, USA
    Augusto
    Wine Packaging

**VIANSA WINERY**
The unifying theme of a collection of 35 wine and food products developed for Viansa Winery is largely derived from classical sources. Renaissance-style frescoes painted on the winery walls inspired the richly colorful yet clean and contemporary look. I wanted visitors to Viansa's tasting room marketplace to feel able to take home the essence of the winery with every purchase.

22  Viansa Winery, USA
    Thalia
    Wine Packaging

23  Viansa Winery, USA
    Reserve
    Wine Packaging

24  Viansa Winery, USA
    Obsidian
    Wine Packaging

**DISTINCTIVE WINE PACKAGING DESIGN THAT INFORMS, ENHANCES AND EXCITES**

# QUON / DESIGNATION
## USA

California native Mike Quon is an artist and graphic designer who is best known for his bold and colorful approach to design.
After coming east to New York, he opened up Mike Quon/Designation to service advertising agencies, publishing companies, and corporations directly.

Much of Quon's approach to design is vividly spontaneous, simple and powerful. He is the son of BBDO art director, package designer and Walt Disney animator Milton Quon, who worked on the classics Fantasia, Dumbo and Bambi. Mike was heavily influenced by his student days when he discovered the joys of the Chinese brush and was also greatly inspired by the Pop Art movement of the 1960s.
His illustrations and designs have been used to help adorn packaging used in diverse arenas, such as health and beauty, breakfast cereal boxes, and both kids' and adults' toys and games.
Mike and his team enjoy using calligraphy and hand lettering, as well as attractively stylish icons, photo illustrations, or collages to express a special tone to each project. Quon's artwork can be found in the permanent collections of the Library of Congress, The New York Historical Society, Wakita Design Museum (Japan), The New York Times, and the US Air Force Art Collection. Mike is the author of two books, Corporate Graphics and Non-Traditional Design, both published by PBC International.

**Mike Quon**
Designation Inc.

tel 732.212.9200
fax 732.212.9217
studio@quondesign.com
www.quondesign.com

543 River Road
Fair Haven, NJ 07704
USA

1  1. Summer in Sicily/Olive Garden
   2. Restaurant Promotion
   3. Promotional Materials
   4. Paper Based Direct Mail
   5. Food and Entertainment

2  1. Nancy & Pengs
   2. Decorative ceramic tiles
   3. Home decor
   4. Pottery & ceramic
   5. House & home/decor

**3**
1. Chills Daiquiri Mix
2. Drink Mix
3. Proposed Product Launch
4. Printed Label
5. Adult Alcoholic Beverage

**4**
1. Pictionary Board Game/Proposed
2. Board Game by Hasbro
3. Entertainment
4. Printed Box and Game
5. Family Home Entertainment

**5**
1. Scents for Urban Outfitters
2. Illustrated label design
3. Perfume & Beauty
4. Label design
5. Cosmetics

**6**
1. Sweet Delights Brand
2. Cologne Spray for Urban Outfitters
3. New product
4. Label design
5. Health & Beauty

3

4

5

6

**7**
1. Fruity
2. Yogurt
3. Food
4. Cardboard Container
5. Convenience Packaging

**8**
1. FTD/Flower Delivery
2. Logo/Symbol for FTD
3. Floral service
4. Brand identity
5. Logo symbol for FTD

**9**
1. Avanti Card Company
2. Gift item/Coffee mugs
3. Greeting cards & gifts
4. Ceramic
5. Set of photographic mugs

**10**
1. Tangram Buddies
2. Proposed book for Blue Apple
3. Book Packaging
4. Cardboard Stock
5. Education & Game

**11**
1. Entertainment Weekly/Time Warner
2. Compact disc package
3. Music Promotion
4. CD direct mail
5. Entertainment & Gossip

**12**
1. Satietrim Energy(with Creative Angle)
2. Heath & fitness drink
3. Personal training
4. Aluminum can
5. Energy & weight loss drink

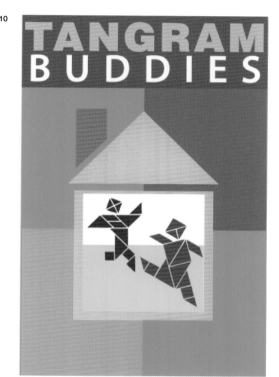

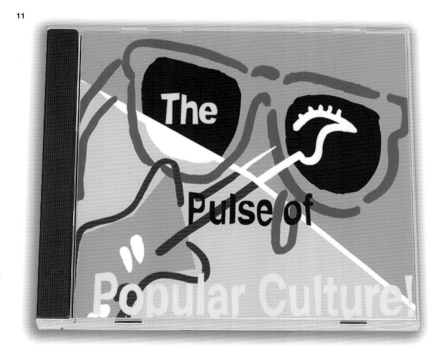

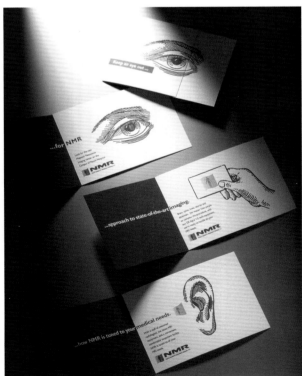

13
1. Infusium 23 / Clairol
2. Hair and Beauty
3. Shampoo
4. Logo for packaging
5. Hair and Beauty

14
1. Sea Breeze for Clairol
2. Skin care and beauty
3. Skin toner & cleanser
4. Silk screen on plastic
5. Health & beauty

15
1. PMC 07/Jimmy Fund
2. Bike race for cancer cure
3. Jersey & shorts with logo
4. Nylon uniforms
5. Annual bike race

16
1. NMR/Nyack Hospital
2. Nyack Magnetic Resonance
3. Direct mail brand identity introduction
4. Cover stock
5. Brochure system/ Direct mail campaign

17  1. Galaxie Pictures/Perdido
    2. Logo for jet
    3. Logo for entertainment industry
    4. Screen printing
    5. Signage for jet

18  1. Bristol-Myers Squibb
    2. Aprovel/Medication
    3. New product introduction
    4. Print & electronic usage
    5. International product launch

19  1. Pepsico
    2. Fast Break promotion
    3. Special promotion for soft drink
    4. Aluminum cans & plastic bottles
    5. Soft drink promotion

20  1. NY Open Center
    2. Institute brochure series
    3. Psychological & physical health
    4. Color coded system
    5. Catalogs for various non profit institutes

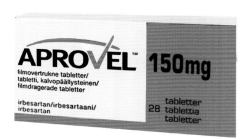

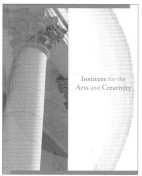

# arbolbranding | Design Studio
MENDOZA, ARGENTINA

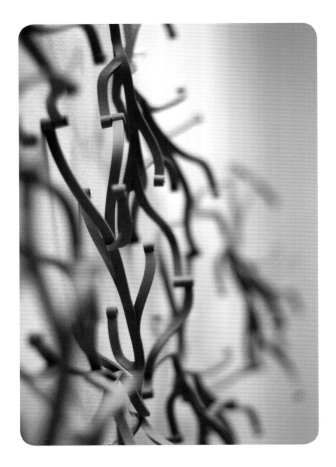

arbolBRANDING is a graphic design studio specialized in visual communication, branding and packaging design.

Our work includes the concept, the strategies of communication and the development of the visual identity of the product, and we follow the production process until the product is finished.

We work with the collaboration of 3D designers, mock-up especialists, marketing experts, illustrators, photographers and translators obtaining creative results for our clients.

www.arbolbranding.com.ar

**arbolBRANDING**
Martínez de Rosas 1040
5500 | Mendoza
Argentina
T: +54 261 429 6830
info@arbolbranding.com.ar
www.arbolbranding.com.ar

1  Monteviejo Winery, Argentina
   JAZZ de Monteviejo
   Packaging Design.

2  Mauricio Lorca Winery, Mendoza, Argentina
   Lorca Lírico Malbec
   Brand Packaging Design.

183

Chupa Chups, Spain
Chupa Chups Relax Mini
Brand Packaging Design in collaboration with EGG-Design Barcelona.

4   Industrias Matas, Argentina
    Brand Packaging Bongú products
    Brand Packaging Design

5   Industrias Matas, Argentina
    Red Cross help food packs
    Brand Packaging Design

6   Finca de Miquilo Winery, Argentina
    Onice Wines
    Branding and Packaging Design

185

# STUDIO 360
SLOVENIA

Vladan Srdic, partner and creative director

Studio 360 is a company providing integrated solutions in the fields of advertising, branding and architecture. The combination of two- and three-dimensional expertise contributes to strategic brand development, efficient results and client satisfaction.

We think 360: branding department offers comprehensive solutions for advertising, illustration, packaging, graphic and web design. Our philosophy is communication in a fresh and sophisticated style: simple, clever and always with a twist. We are focused on efficient solutions and dedicated to obtaining a higher value for your brand - leaving nothing to chance. Every project is the most important to us- combining the principles of aesthetics, idea and function, we are commited to creating strong concepts and added value for our clients.

We do 360: architectural department focuses on flexible design concepts and innovative building techniques that improve the quality of living. Every building receives a comprehensive solution: analysis of context, client's brief, investment and construction technologies. We aim to achieve more with less, and creatively transform constraint into opportunity. Studio 360 provides services encompassing, but not limited to, urban planning, exhibition set-up, building exteriors and interiors. Design is based on simplicity, attention to detail and intelligent use of materials.

Studio 360 have been featured in distinguished books and magazines and have won many awards at international competitions. We run all of our projects, many of them for clients around the world, from our Ljubljana based studio.

**Studio 360 d.o.o.**

Address : Kotnikova 34, SI-1000 Ljubljana, Slovenia
Phone / Fax : +386 (0)1 431 3312
Mobile : +386 (0)31 847 261
E-mail : office@studio360.si
Web page : www.studio360.si

Cilent : Peter Wegele
Country : Germany
Year : 2008
Brand : Necessarily Two
Type of product : CD cover

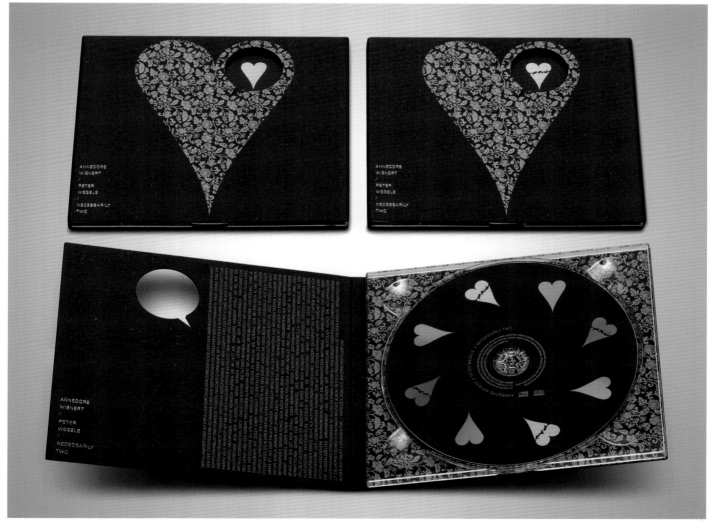

# THE CREATIVE METHOD
## AUSTRALIA

Left to right: Andi Yanto, designer; Pip Hopkins, account manager; Tony Ibbotson, creative director; Mayra Monobe, designer.

The Creative Method has only been in existence since October 2005 and in 3 short years we have been mistaken for a fertility clinic, moved twice, eaten too many takeaways, stolen our neighbours cutlery, lost a lot of pens, drunk a silly amount of alcohol and had a lot of fun.

We have also won a number of design awards, appeared in numerous publications, and made our clients very rich.

Our core focus is creating new-to-world consumer brands and fixing some of the older ones. We work on a large spectrum of clients and projects from global companies such as Diageo and Unilever to small family businesses and yummy cake makers.

Currently our full-time staff members consist of an Indonesian, an Australian, a New Zealander and a Brazillian, so we are equal opportunity employers where a good visual language is our mother tongue.

The aim for the short term is to be one of the best in the world; long term - world peace would be nice.

www.thecreativemethod.com

The Creative Method
Studio 10, 50 Reservoir Street
Surry Hills NSW 2010
Australia
T +61 2 8231 9977
F +61 2 8231 9980
us@thecreativemethod.com
www.thecreativemethod.com

1  Guzman y Gomez, Australia
   Guzman y Gomez Mexican Table Salsa
   Graphic Design
   The Creative Method

2  Potato Magic, Australia
   Sultry Sally Potato Chips
   Graphic Design
   The Creative Method

3 Diageo, Australia
Real McCoy Bourbon Whiskey
Graphic Design
The Creative Method

4 Marlborough Valley Wines, New Zealand
Fire Road Label
Graphic Design
The Creative Method

5 Diageo, Australia
Baileys Gift Box
Graphic Design
The Creative Method

6   Mt. Olympus Wines, New Zealand
    Elbows Bend Label
    Graphic Design
    The Creative Method

7   Saint Clair Family Estate, New Zealand
    Pioneer Block Label
    Graphic Design
    The Creative Method

8   Saint Clair Family Estate, New Zealand
    Saint Clair Single Gift Box
    Graphic Design
    The Creative Method

191

# EICHE, OEHJNE DESIGN
## GERMANY

Most markets are so competitive that the marketing departments of companies must act with utmost flexibility and a highly professional attitude. With our core competence – Corporate Design and Packaging Design – we think that we can offer instruments to meet the demands of the market. We believe Corporate Design and Packaging Design are an extension of brand development. For many people, the word "corporate" describes a static system – one that is inflexible and difficult to change. We believe, however, that corporations are dynamic systems. This defines our approach to the market and design. That's how corporations stay flexible and fast-moving.

EOD offers a full range of services to execute Corporate Design and Packaging Design projects, and these services come with personal involvement and innovative thinking. This includes consulting, strategy and conceptualization – from design to the end product. In the process of designing and implementing projects, we are always striving for distinctive, progressive and high-quality solutions that distinguish our clients from the masses. In short – our goal is to combine systematic thinking with first-class design. We have been rewarded for our approach with long-term business relationships and international prizes. Some of our leading ideas are:

· Steer toward a unique, brand-typical appearance on the market
· Increase visibility via consistent and progressive brand design
· Raise customer loyalty through precise definition of the target audience
· Improve turnover through high-quality design
· Raise the value of the company through increased brand value

Since we were founded in 1999, we have worked for numerous companies, institutions and associations. Both mid-sized service providers and large corporations.

Ilka Eiche and Peter Oehjne, founder and owner of the agency

Fritz Allendorf Winery, The Premium Line

EICHE, OEHJNE DESIGN / GERMANY

FRITZ ALLENDORF WINERY
THE PREMIUM LINE
WHITE WINE
GLASS BOTTLE AND PAPER LABEL
PREMIUM STRATEGY
STRUCTURAL PACKAGING &
BRAND IDENTITY

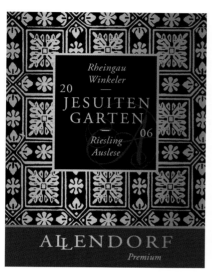

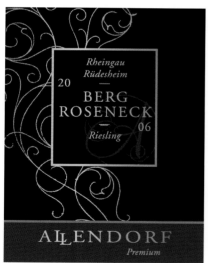

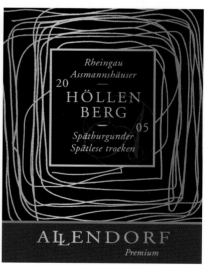

FRITZ ALLENDORF WINERY
THE PREMIUM LINE
WHITE WINE
DIFFERENT PAPER LABELS
PREMIUM STRATEGY
STRUCTURAL PACKAGING &
BRAND IDENTITY

EICHE, OEHJNE DESIGN / GERMANY

FRITZ ALLENDORF WINERY
LINE EXTENSION WITH OWN BRANDING
WHITE WINE
GLASS BOTTLE AND PAPER LABEL
LIFESTYLE STRATEGY

ECCO TERRA
QUINOA CEREAL PRODUCTS
CELLOPHANE WRAP &
PAPER LABEL
HERITAGE STRATEGY

# DENEY DESIGN
## TURKEY

Murat Celep, founder and creative director

Deney has been established in 1997 as a boutique design office. We have been providing high-quality corporate identity and packaging design services.

We, at Deney Design, start every design project with research. This issue is very important for us. Information provided through research is the key-point which will lead us to the result. We believe that every creative designed should be backed with a good idea. All our designs reflect such process. Although we provide design services in a broad portfolio of business areas, corporate identity and packaging design has always been of special importance for us. In time, we have had sufficient experience for gaining expertise in these two areas.

It is one of the most important issues of the package brand wars. When we handle the package itself as an advertising material, the campaign would be much more effective with a story and a viable remark. Our aim is designing persuading packages with a character and which can be easily differentiated at the locations where purchasing decisions are made.

We have cooperated with a lot of large and small companies providing services in food, medicine, construction, electronics, textile, iron and steel, publishing, cosmetic, hard goods, furniture, telecommunication, robotics and automation, electric home appliances, education and marine industries. We have provided services for global brands with big names and small companies manufacturing in small plants. We can say that these companies of various sizes had something in common; they attached importance to design. We experienced the way how creative designs provided added value and competitive power for lots of small brands.

**www.deney.com.tr**

Deney Design
Ismail Pasa Sokak 70/1 Kosuyolu
34718 Istanbul - Turkey
T : +90 (216) 428 79 97
F : +90 (216) 428 79 98
deney@deney.com.tr
www.deney.com.tr

1 - 2. Murpa Tea
　　 Black Teabags
　　 Graphic Design

3. Murpa Tea
　 Bergamot Flavoured Tea
　 Graphic Design

4. Murpa Tea
   Bergamot Flavoured Tea
   Graphic Design

5. Agrobest Food
   Gourmed Olive
   Graphic Design

6. Anatolia Wines
Dry Red Wines
Graphic Design

# TRIPLE 888 STUDIOS
## AUSTRALIA

Incorp. **TRIPLE 888 INTERNATIONAL**
Design And Marketing Solutions

**Head Office** 81-83 Wigram Street
Parramatta NSW Australia 2150
**p** +61 2 9891 2888
**f** +61 2 9891 1283
**e** designit@triple888.com.au
**www.triple888international.com**

Established in 1986, Triple 888 Studios/Triple 888 International has grown to provide a very broad range of creative design and strategic marketing services.

Our strong design team have produced many award winning creative solutions to clients needs – from packaging (import and export), brochures, corporate images, displays, advertisements and web sites. Our services also include Marketing Strategies, Media Planning, Public Relations, Direct Marketing, Event Management, Television and Radio Production, Sales Promotions and Database management for specialised clients.

Triple 888 Studios/Triple 888 International provide creative artwork services of the highest standards with meticulous attention to detail, showcasing each client's products in the best way possible. As a successful business, our services are employed by local and international clientele and has prepared language specific artwork for Europe, United Kingdom, Asia Pacific, South Africa and Gulf regions.

Clients consist of manufacturers operating in a wide range of industries including pharmaceutical, industrial, housing, cosmetics, automotive, homewares and consumables. These clients include GlaxoSmithKline, BP-Castrol, DEB Australia, Felton Homes, Royal Australian Navy, Sheldon and Hammond, Jonsa Ellies, Aspen, Miller's Retail, Southern Cross, Le Mac Enterprises, Form-Tek, Universal Publications, PT Hydraulics Australia, Form-Rite & Markets Unlocked.

We maintain a strong consultative business partnership with our clients, with the objective of always exceeding their expectations. This flair for innovation has lead to several other awards recognising excellence. These include:

- National Print Awards
- Australian Catalogue Awards
- Australian Packaging Awards
- Summit Creative Awards
- Supplier Recognition Awards
- Premier Print Awards (worldwide)
- Creative Awards
- Western Sydney Arts Business Awards

We constantly strive to achieve the highest standards of services and in doing so, have created long lasting partnerships with our clients.

1 Papaya
  Homeware

2 Aussia
  She-Hydro

3   Sheldon and Hammond
    Avanti, Sierra Knife block
4   Sheldon and Hammond
    Avanti, Vista Knife block
5   Sheldon and Hammond
    Mundial, Grande
6   GlaxoSmithKline
    Vaccine Packs
7   The Le Mac Australia Group
    Christmas Promotion Bottle

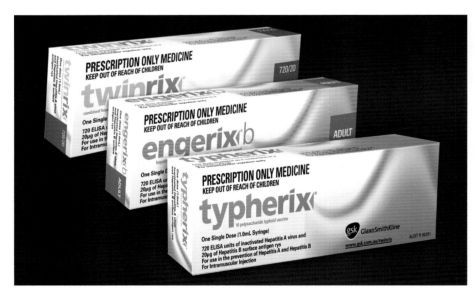

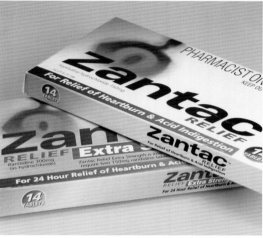

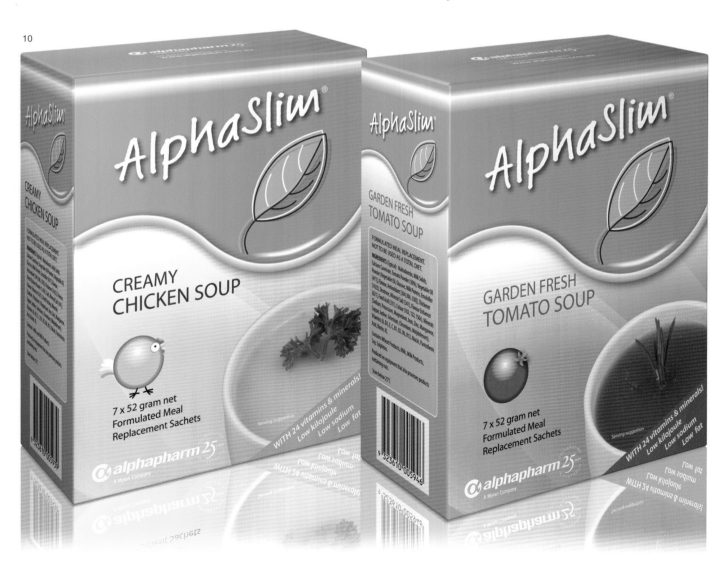

8 Select Health
  Select Lifestyle,
  Vitamin range
9 GlaxoSmithKline
  Zantac relief
10 Alphapharm
   AlphaSlim Soup

# MINIRA CREATIVE
## JAPAN

**MINIRA CREATIVE**

Eiji Nakao

Founded on August 8, 1996, our company specializes in alcoholic beverage-related projects-simply because everyone at the company loves to drink.
But the biggest drinker is our president who is an SSI-authorized sake sommelier.
We deal with both foreign liquor and Japanese sake, and we're also good with food projects that pretty much go hand-in-hand with drinking.
Our goal is to become a "gourmand planning and production company" for all of your palate-teasing needs.

**MINIRA CREATIVE**
#903 Minami aoyama Uni heights, 6-12-3
Minami aoyama, Minato-ku Tokyo
107-0062, JAPAN
TEL +81 3 5468 0878  FAX +81 3 5468 0898
e-nakao@minira.co.jp
www.minira.co.jp

1. Okunomatsu Shuzo
   Okunomatsu/shochu
   label design
2. Oyama shuzo
   Oyama Brand Identity/sake
   label design, calligraphy
3. Oyama shuzo
   Fuin-shu/sake
   package design, label design

4 Urayasubashi Brewery
  Wangan-maihama beer/Beer
  label design

5 Mercian
  Jugemu/shochu
  label design

6 Mercian
  Vintage single cask malt whisky/malt whisky
  label design

7   Kokubu confectionery
    Sweets OGOZYO/confectionery shop
    CI design,package design,logo design,display

8   Roba kashi tsukasa confectionery
    L'HOTEL de KITAKURABU/confectionery shop
    CI design,package design,logo design,display

9   Schick Japan
    Shaving Gel/shaving gel
    package design

10  Schick Japan
    Shave guard/shaving foam
    package design,illustration

# NARROW HOUSE
SPAIN

At Narrow House we believe in listening before speaking.

We could talk about the importance of branding and marketing and spend hours listening to the sound of our own voices, but that's not really our style. Instead, we prefer to listen. We listen to our clients, and we listen to their customers before drawing any conclusions, and once we start work our designs tend to speak for themselves… and what they say is often quite interesting!!

1.

## narrow house
Branding, Packaging & Diseño Estratégico

www.narrow-house.com

**1. Client:** Helados Royne. **Country:** Spain. **Brand:** Bigbom. **Type of Product:** Ice cream. **Material:** Plastic wrap. **Theme:** "SmmmMooothh chocolate ice cream"
**2. Client:** Olivos Clavero. **Country:** Spain. **Brand:** Clavero Extra Virgen. **Type of Product:** Olive oil. **Material:** Adhesive label / Glass. **Theme:** "extra special olive oil"
**3. Client:** Pedro LLinares. **Country:** Spain. **Brand:** Jamón Extra Fino. **Type of Product:** Cured ham. **Material:** Plastic pack. **Theme:** "a Spanish delicacy"
**4. Client:** Helados Royne. **Country:** Spain. **Brand:** Maracay. **Type of Product:** Ice cream. **Material:** Plastic wrap. **Theme:** "everybody loves ice cream"
**5. Client:** AWO Hoteles. **Country:** International. **Brand:** Base. **Type of Product:** Toiletries. **Material:** Plastic tubes. **Theme:** "a clean design"
**6. Client:** Bodegas Luque. **Country:** Spain. **Brand:** Amontillado 1888. **Type of Product:** Wine. **Material:** Adhesive label / Glass. **Theme:** "tradition and origin"

**7. Client:** Helados Royne. **Country:** Spain. **Brand:** Roymilk. **Type of Product:** Ice cream for kids. **Material:** Cardboard. **Theme:** "ice cream is fun"

**8. Client:** G&B Restaurant and Lounge. **Country:** UK. **Brand:** G&B House wine. **Type of Product:** Wine. **Material:** Adhesive label / Glass. **Theme:** "wine that makes you smile"

**9. Client:** Pasión for Fashion www.pasionforfashion.com. **Country:** Spain / International. **Brand:** Pasión for Fashion. **Type of Product:** Fashion Boxes. **Material:** Cardboard. **Theme:** "limited editions"

# CINQ DIRECTIONS INC.
JAPAN

eas are package design, advertising design, interactive design, and so on.

We think like this.
Recently many designers are placing their main emphasis on the points of "How does it sell", and "How is it consumed ".
However, we don't think these are the only important points.
We believe that good design should be about its "Relationship with the person".
This means that it is necessary for the design to fit into society correctly.
The design should exist in human-life relevantly, and according to society's needs.
To achieve this goal, we are always researching the relationship between the client and the consumer.
We continuously endeavor to create a design that fits the social relevance of the consuming society.
Our primary goal is that we must always practice this concept.
We believe this goal can be successfully accomplished, through the power of good design.

**http://www.39d.co.jp**

CINQ DIRECTIONS INC.
301 Altis-Akasaka
8.5.2 Akasaka Minato-Ku.
Tokyo. 107.0052. Japan
Tel. 03.3405.7586  Fax. 03.3405.7592
URL http://www.39d.co.jp
E-mail hoso@39d.co.jp

1 DyDo DRINCO
POCKET JUICER STAND
FRUIT JUICE
ALUMINUME
DESIGN TO EXPRESS FRESHNESS

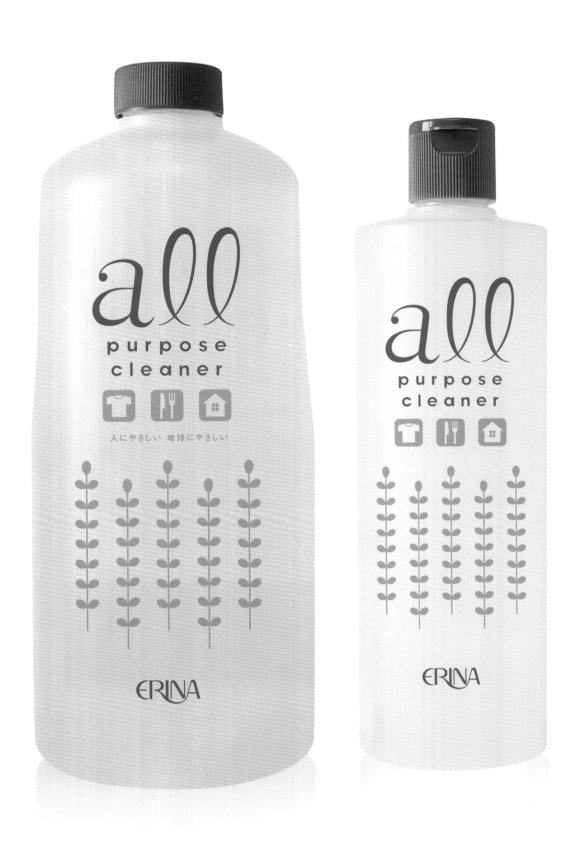

2 ERINA / Japan
ALL PURPOSE CLEANER
CLEANER
PIASTIC
DESIGN TO EXPRESS NATURAL ORIGIN

3  Tamanoi / Japan
Diet Tamanoi
APPLE VINEGER
GLASS/PAPER
DESIGN TO EXPRESS STRONG ATTENTION

# Bouzón | Comunicación y Diseño

ARGENTINA

## Bouzón | Comunicación y Diseño

Bouzón | Comunicación y Diseño is a consulting company specialized in Corporate Identity and product Image.

The designer and university professor Alejandro Bouzón, who has been in the market since 1993, leads its team of professionals and provides services of advice, planning, creative development and implementation of communication and design strategies.

Aiming at optimizing the brand image as well as the product in the design of packaging and guaranteeing the top performance of each investment made on design, each and every project is approached as part of a whole, beginning by the analysis and diagnosis of the case at the issue, the mapping of a sound estrategic plan and a consequent creative process to explore the design alternatives that best adjust to the communication objectives and the commercial e strategic of the client.

Contributing with their knowledge and experience, intervention in each design is implemented with high fidelity and a utmost utilization of material, technological and economic resources. By adding creativity and effectiveness to communication and products, Bouzón | Comunicación y Diseño carries out its work methodology by the side of their national and multinational customers, some of which are mentioned below. Osram, Siemens, AFIP, Banghó, Dapsa, CPAU, Totalgaz, Cervi, Total, Jumbo, Alice, Roemmers.

www.bouzon.com.ar

Bouzón | Comunicación y Diseño
11 de Septiembre 4237 8° of B
C1429CJB | Buenos Aires
Argentina
T/F +54 11 4702 7197
info@bouzon.com.ar
www.bouzon.com.ar

1   Banghó S.A. | Argentina
    Banghó®
    Personal computers
    Mounted cloaked pasteboard in corrugated cardboard reinforced
    Packaging graphic design and trademark design

2 Mario Cervi e Hijos S.A. | Argentina
Cervi
Fresh fruits
Corrugated cardboard
Packaging graphic design and trademark design

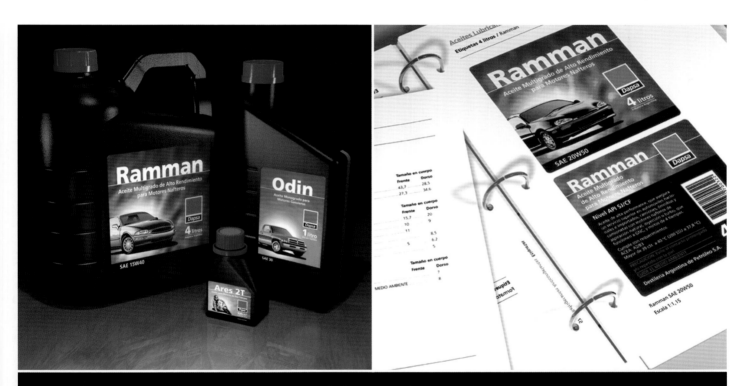

3   Destilería Argentina de Petróleo S.A. | Argentina
    DAPSA
    Oils lubricants
    OPP labels
    Graphic design system of labels

4   Agosti Hnos. S.A. | Argentina
Arroz Dos Hermanos
Line of rices
OPP labels
Labelsgraphicdesign system and trademark design

5   Nantrat S.A. | Argentina
Saint Germain
Line of accessories for wine
Role illustration with rolling of OPP kill
Graphic design system of labelsand redesign of trademark

# THE KITCHEN COLLABORATIVE
USA

Aniko Hill, Principal and Creative Director
*All Product and Portrait Photography by Jesse Hill*

We're not a real kitchen, we don't even work in one — although we do like to break for food from time to time. Our name is purely a metaphor — but one that we think describes what we do and how we do it.

The Kitchen Collaborative is an award-winning boutique branding agency that offers a unique specialty focus on the female consumer market. Our agency was born from a strong desire to create an alternative to the typical top-heavy design studio. When you take away the politics, titles and extravagant overheads, what's left? Creative agility. A leaner, more efficient process. The ability to harness the best talent for each challenge. And a focus on creating quality branding and design services that are individually tailored to meet each and every client's goals.

Whether we're working on a comprehensive brand strategy, collateral or advertising campaign, packaging system, space design or website, we work smart and we do smart work. We may not be a real kitchen — but we do know what it takes to cook up a strong brand.

www.kitchencollaborative.com

The Kitchen Collaborative
Los Angeles, California
United States
T 1 818 588 3060
menu@kitchencollaborative.com
www.kitchencollaborative.com

1   TESS (Teen Everyday Skincare System), United States
    TESS
    Teen skincare line
    PET containers / printed clear labels

2   Curtis Mathes and Brand Texture, United States
    Curtis Mathes
    LCD televisions / digital picture frames
    Printed cartons

3   Sea-Yu Enterprises and Brand Texture, United States
    Petrotech
    All natural pet odor eliminator products
    Non-CFC cans / printed labels

4 Make-Up Designory, United States
Cosmetics line
Mixed stock components / printed cartons

5 Agera Laboratories, United States
Agera
Skincare line
Mixed stock components / printed cartons

# DEPOT WPF
## RUSSIA

**Depot WPF Brand and Identity** is a leading Russian independent agency. Founded in 1998.

**Services**
Our core competences are consumer and corporate branding. We provide integrated solutions including brand strategy, brand visual identity and corporate identity.
Our services include development of brand positioning, communication strategy, naming and positioning slogan creation, brand visual identity, corporate identity, brand guidelines, packaging design, POSM and promo-materials creation.

**Clients**
Depot WPF has been working with leading Russian and international companies who aim to take or strengthen their position on local market.
Clients: Danone, TOTAL, Nestle Foods, Unilever, Europe Foods G.B., Myllyn Paras Oy, Multon, Unimilk, Beeline, and other.

**Awards**
Depot WPF is the winner of different awards such as EPICA AWARDS 2006, EPICA AWARDS 2007, Golden Drum 2006, finalist of Golden Drum 2007, Golden Hammer 2008, Kiev International Advertising festival 2000, 2006-2008, Moscow International Advertising Festival 1998-2008, EFFIE / BEST BRAND'2000, EFFIE / BEST BRAND'2002.

**Contacts**
Russia 109004, Moscow,
Pestovsky pereulok, 16, bld. 2
Bussines-Center AKMA
Tel./Fax +7 (495) 363-2288
Email: info@depotwpf.ru
Web: www.depotwpf.ru

1. **Grossmart, Russia,**
**Memoire,**
Cognac,
Depot WPF Brand&Identity

1. Grossmart / Cognac

2. Nestle / Confectionery

2. **Nestle, Russia, Comilfo Gold,** Confectionery, Depot WPF Brand&Identity

3. **Nestle, Russia, Comilfo post card,** Confectionery, Depot WPF Brand&Identity

3. Nestle / Confectionery

4. Sady Pridoniya / Juice

4. **Sady Pridoniya, Russia,
Zolotaya Rus,**
Juice,
Depot WPF Brand&Identity

5. **Uvelka, Russia,
Italianskaya trapeze,**
Rizotto,
Depot WPF Brand&Identity

6. **Mironov, Russia,
Russkie pelmeni,**
Ravioli (pelmeni),
Depot WPF Brand&Identity

5. Uvelka / Rizotto

6. Mironov / Ravioli (pelmeni)

6. Severnaya kompaniya / Fish

6. **Severnaya kompaniya, Russia, Moremaniya,**
Fish,
Depot WPF Brand&Identity

7. **Agropromishlenniy park, Russia, Vegetoriya,**
Packed vegetables,
Depot WPF Brand&Identity

7. Agropromishlenniy park / Packed vegetables

8. French laboratory of BIOGRAPHY OF BEAUTY / Cosmetics

9. Pharmaceutical chain 36'6 / Cosmetics

8. **French laboratory of BIOGRAPHY OF BEAUTY, Biography,**
Cosmetics,
Depot WPF Brand&Identity

9. **Pharmaceutical chain 36'6, Russia, NaturAge,**
Cosmetics, Depot WPF Brand&Identity

10 **Pharmaceutical chain 36'6, Russia, Dobraya zabota,**
Baby care products,
Depot WPF Brand&Identity

10. Pharmaceutical chain 36'6 / Baby care products

11. Yaroslavskie kraski / Paints

11. **Yaroslavskie kraski, Russia, Premia,**
Paints,
Depot WPF Brand&Identity

12. **Hygiene Kinetics, Russia, Ola!,**
Personal hygiene products
Depot WPF Brand&Identity

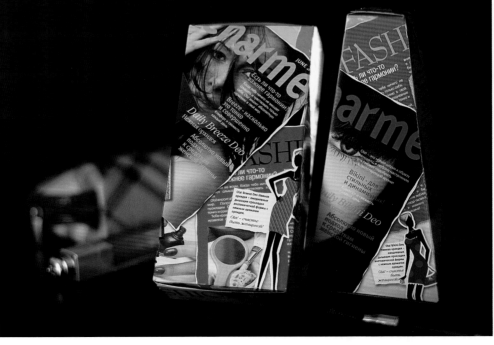

12. Hygiene Kinetics / Personal hygiene products

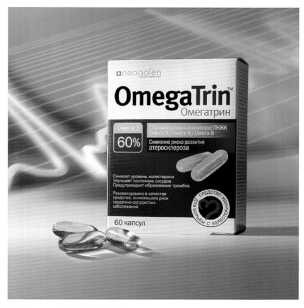

13. Newman nutrients AG / B.A.D.

14. Pharmaceutical chain 36'6 / Cosmetics

13. **Newman nutrients AG, Russia -Switzerland, Omegatrin,**
B.A.D.,
Depot WPF Brand&Identity

14. **Pharmaceutical chain 36'6, Russia, NaturAge,**
Cosmetics,
Depot WPF Brand&Identity

15. **Newman nutrients AG, Russia-Switzerland, SlimCode,**
B.A.D.,
Depot WPF Brand&Identity

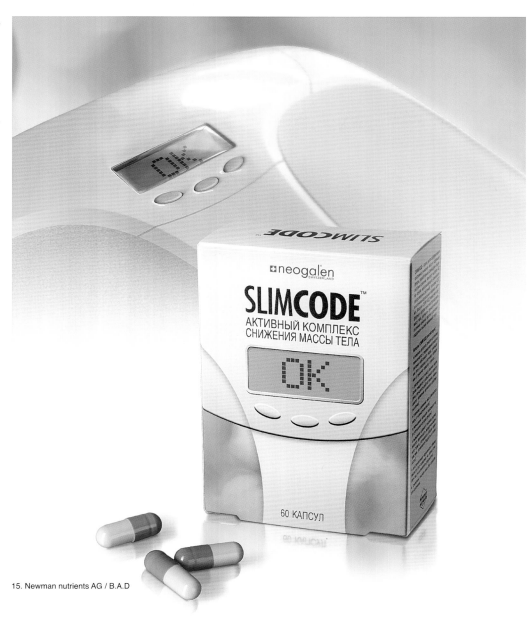

15. Newman nutrients AG / B.A.D

# SYLVIE DE FRANCE DESIGNER
## FRANCE

Sylvie de France, Designer
29 rue Miguel Hidalgo
75019 | Paris
France
T/F+33 1 42 38 37 99
Sylviedefrance@sylviedefrance.com
www.sylviedefrance.com
Contact : Gwenola Garel
Marketing & Sales Manager
gwenola.garel@sylviedefrance.com

## APPROACH

• The agency:

Based in Paris, France, the Sylvie de France Designer agency specializes in packaging creation, volume design and graphics for luxury products, since 1997, namely:

Perfumery    -    Cosmetics    -    Fine Spirits

• Sylvie de France:

Sylvie de France has dedicated the last twenty years of her life to the world of luxury products. After graduating from the ESDI school (Ecole Supérieure de Design Industriel) and the École des Beaux-Arts (Fine Arts Paris), she was soon working for the most famous international luxury brands.
Her method: she finds it indispensable to work closely with clients in order to decode their key values and remain faithful to their roots, image and needs.
Attentiveness, client relations and the exchange of ideas are essential elements of Sylvie de France's approach to designing high-quality products. In addition to beauty and harmony, her designs aim to preserve the very soul of the brand with sensitivity and a touch of unexpected.
Indeed, Sylvie de France brings to her work a high-level of creativeness and an unquestionable technical expertise. Over the years she has developed many technical innovations for the likes of Lolita Lempicka, Jean-Charles de Castelbajac and Ungaro.

## EXPERTISE

The Sylvie de France designer agency accompanies brand development from start to finish:
- Volume design: bottle creation, accessories, technical follow up of the project.
- Graphics: visual identity, packaging (cases, boxes).
- Publishing and Product Communication: press files, marketing books, shopping bags, gifts, jewels, samples, blotters, T-shirts.
- Merchandising: POS, furniture.
- Consulting in brand strategy: naming, brand image, trends, colour and material research.

## CUSTOMERS REFERENCES

AGATHA RUIZ DE LA PRADA - CACHAREL - CAMUS COGNAC - CASTELBAJAC - CHAMPAGNE NICOLAS FEUILLATTE - CHANTAL THOMASS - EDEN PARK - EMANUEL UNGARO - GIVENCHY - ISABEL DERROISNE - ISSEY MIYAKE - JULIETTE HAS A GUN - JP GAULTIER - KIOTIS - LACOSTE - LOLITA LEMPICKA - NINA RICCI - OLIVIER STRELLI - SALVATORE FERRAGAMO - STELLA CADENTE - VAN CLEEF & ARPELS - YVES ROCHER

www.sylviedefrance.com

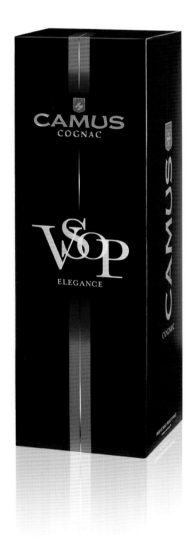
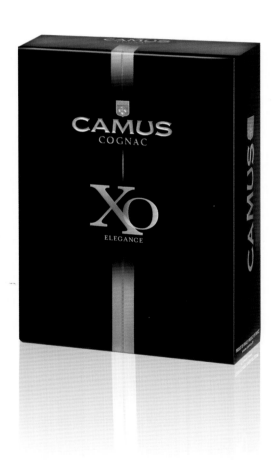

1  Camus Cognac | France
   VS, VSOP, XO
   Cognac
   Creation of a New Graphic Identity

2  Procter & Gamble | Global
   Lacoste Dream of Pink
   Fragrance
   Volume and Graphic Design

3  Parfum Ferragamo | Global
   F for Fascinating
   Fragrance
   Volume and Graphic Design

4  Parfums Lolita Lempicka | France
   Fleur Défendue
   Fragrance
   Volume and Graphic Design

5  Parfums Givenchy | Global
   Ptisenbon in the woods
   Fragrance
   Graphic Design

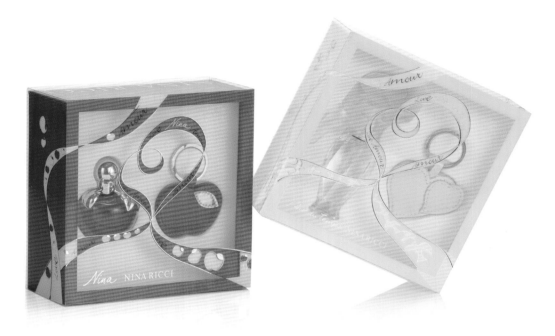

6   Parfums Nina Ricci | Global
    Nina and L' Air du Temps
    Fragrance ( Gift Set )
    Graphic Design

7   Ungaro Parfums | Global
    Ungaro Man
    Fragrance for men
    Volume and Graphic Design

8   Champagne Nicolas Feuillatte | Global
    Magie Baltique
    Graphic Design

# KIRK WYLAM DESIGN
### ENGLAND

My philosophy is simple, produce great design and the client will be happy, as the results should speak for themselves. I believe successful design is achieved by understanding the client's objectives and by being passionate about design and the benefits it can bring to a company.

**Kirk Wylam Design**
49 Downing Street, Farnham,
Surrey GU9 7PH England

*t:* +44 (0) 1252 821838

*e:* roywylam@kirkwylam.com

*w:* www.kirkwylam.com

**Client:** Greenman Group PLC
**Product:** Recycled Toner Cartridge
**Product Range:** Four sizes

### Design Brief

Produce a design approach that would allow the user to recognise that this product is a credible alternative to buying new.

For sale through specialist retailers, the design needed to position Access as an 'Original Alternative' as well as a cost-effective alternative to buying new, enabling the retailer's to maintain sensible selling prices and margins and the user to have a quality product.

The technical aspect of the brief was the packaging was going to be printed by Flexography in four colours, directly on to cardboard; this process has a 3mm register variable in the printing.

### Design Solution

**Brand**

Design and creation of the brand used a Zebra motif to emphasize the black toner and to have an environmental feel; the access name is designed to be a punchy modern logotype with shelf presence.

**Packaging**

The design is striking, positioning the brand as an 'Original Alternative'. Using a multi-colored moving effect creates a distinctive and instantly recognisable pack.

This was followed up with the strap line on the bottom of the pack clearly setting out what the products position is.

**Technical**

Black was used to key the printing process, making sure there is no out of register printing.

Left to right: Ben Kawaichi, Jamie Capozzi, Marwan Salfiti

Theory Associates entrance

# THEORY ASSOCIATES

Theory Associates specializes in targeted identity systems and product packaging for the consumer electronics market.

Located in the SOMA district of San Francisco, Theory Associates houses a tight knit group of designers, illustrators, and writers who share a common addiction: technology. Or more accurately: gadgets. The bustling, downtown atmosphere energizes creativity, inspires ideas, and keeps us close to the people who matter most—customers.

We approach each job with a basic process that involves listening, questioning, comparing, and analyzing to develop a sound plan to achieve the client's goals. This strategy provides the direction we need to execute initiatives with engineering-like precision. The objective is always the same—help a company connect with its customers. Achieve this and everything else will follow.

**TheoryAssociates.com**

Theory Associates
300 Brannan St, Suite 503
San Francisco, CA 94107
USA
T +1 415 904 0995
info@theoryassociates.com
www.theoryassociates.com

1  Monster Cable, Inc, USA
   Monster Performance Car
   Industrial and Graphic Design

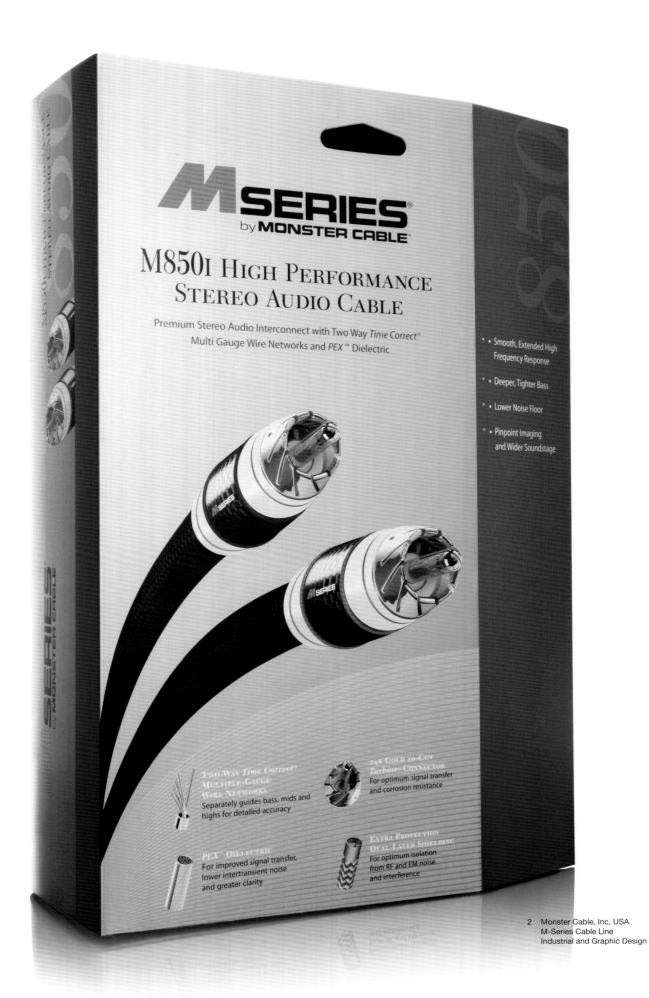

2  Monster Cable, Inc, USA
M-Series Cable Line
Industrial and Graphic Design

3   Monster Cable, Inc, USA
    Monster Game XBox 360 Cable
    Industrial and Graphic Design

4   Monster Cable, Inc, USA
    Monster THX Cable Line
    Industrial and Graphic Design

# SABINGRAFIK, INC.
## USA

Tracy Sabin, designer and illustrator

Sabingrafik, Inc. provides graphic design services with an illustrator's sensibility. This small firm produces unique package designs that feature a strong pictorial element. All aspects of the design, from the wrapping paper stock, to the custom typography, to the choice of color palette, to the illustration concept and style are considered in order to create a unique and memorable solution.

**tracy.sabin.com**

1   Seafarer Baking Company | United States
    Spanky Chocolate Hazelnut Bar
    Baked Goods
    Paper wrapping / Paper labels
    Packaging & label design

Sabingrafik, Inc.
7333 Seafarer Place
Carlsbad, CA 92011
USA
T 760 431 0439
tracy@sabin.com
tracy.sabin.com

2  Seafarer Baking Company | United States
Biscotti al Cioccolato
Baked Goods
Paper wrapping / Paper labels
Packaging & label design

3 Seafarer Baking Company | United States
Gateau Breton
Baked Goods
Paper wrapping / Paper labels
Packaging & label design

4  Seafarer Baking Company | United States
Panforte al Cioccolato
Baked Goods
Paper wrapping / Paper labels
Packaging & label design

5  Seafarer Baking Company | United States
   Turkish Delight
   Baked Goods
   Paper wrapping / Paper labels
   Packaging & label design

6   Seafarer Baking Company | United States
    Panettone
    Baked Goods
    Paper wrapping / Paper labels
    Packaging & label design

**BRANDING.IDENTITY.DESIGN**
8 Bakopoulou St., GR-15451 N. Psichiko  T. +30 210 67.53.740, F. +30 210 67.23.044
info@milk.com.gr, www.milk.com.gr

1. 
   1. G.S. Skiadaresis, Greece
   2. Skiadaresis
   3. Traditional Greek Delights
   4. Carton Box / Foil / Bag
   5. Branding and Packaging design for traditional delights / Milk Ltd.

2.

2. 1. 3E (Hellenic Bottle Company), Greece
   2. Frulite On the Go
   3. Fruit drink
   4. PET
   5. Branding and Packaging design for fruit drink/ Milk Ltd.

3. 1. 3E (Hellenic Bottle Company), Greece
   2. Avra Bloom
   3. Bottled water
   4. PET
   5. Branding and Packaging design for children bottled water/ Milk Ltd.

3.

4. 1. Lavipharm, Greece
   2. Castalia
   3. Dermocosmetics
   4. Carton Box & PE (polyethylene)
   5. Branding and Packaging design for dermocosmetics / Milk Ltd.

4.

5.

5. 1. Nestle, Greece
   2. Boss
   3. Ice Cream
   4. Foil
   5. Evolutionary rebranding / packaging for premium ice cream brand/ Milk Ltd.

# VIE DESIGN STUDIO
## INDONESIA

**Hendra Wijaya - Senior Graphic Designer**

In the year of 2002 in Bandung City-Indonesia was a start of our business in the area of Creative Branding and Strategic Communication.

Initiated from our passion in creating design that's effective, communicative with high artistic value, we try to accomodate our client's need of maximum result. In the process of creating a design, we always try to do a personal approach because we believe that the key to success of a design is to understand and to learn our client's business.

Other than that, we also invite our client to participate and to enjoy in the process of creativity, aligning our minds through the design that we created.
By then, perception about a good design will be understood by our client and we expect that our client could give a good appreciation of the art of communication.

Vie design offers an integrated solution which includes: coorporate & retail identity, branding & packaging, annual report, company profile, digital imaging, environmental graphic design.

**CONTACT**
**Vie Design Studio**
a: Ciateul Tengah 7, Bandung-INDONESIA
e: azrael@bdg.centrin.net.id
   viedesign@yahoo.co.id

1. Client Name: Dildos Assorted SL(SPAIN)
   Brand Name: Zahara
   Label & Gift Box
   Material: Art Paper
   Theme: Ethnic, Sensual & Romantic

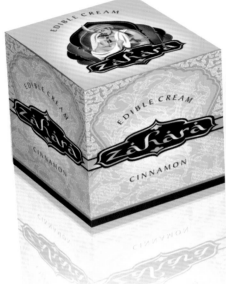

2. Client Name: Dildos Assorted SL (SPAIN)
   Brand Name: Zahara
   Label
   Material: Art Paper
   Theme: Ethnic, Sensual & Romantic

3. Client Name: Sharon Bakery/INDONESIA
   Brand Name: Roomboter
   Material: Plastic

4. Cajna Pasteta (CROATIA)
   Label & Pate

5. Jordan Bakery & Cafe (INDONESIA)
   Material: Plastic

# ZERO DESIGN LIMITED
## SCOTLAND

1. (1) Dent Brewery Limited, UK
   (2) Aviator Ale
   (3) Traditional real ale
   (4) Labelled glass bottle
   (5) One of a range of 6 real ales

2. (1) Fairworld Limited, UK & Brazil
   (2) Organic Brazilian Honey
   (3) Wild honey from the Amazon Jungle
   (4) Labelled glass jar & tag
   (5) One of a range of organic products from Brazil

3. (1) Douglas Laing & Sons Limited, Scotland
   (2) Clan Denny
   (3) Vatted malt whisky from Islay and Speyside
   (4) Labelled glass bottle
   (5) One of two "gift whisky" products

1.

2.

3.

ZERO DESIGN LIMITED  
The Bond Building  
13a Breadalbane Street  
Edinburgh EH6 5JJ  

t: +44 (0)131 554 9930  
f: +44 (0) 131 554 8569  
e: studio@totallyzero.com  
w: www.totallyzero.com  

4. (1) Albyn Limited, UK
   (2) Sunsure
   (3) Sunburn detection lens
   (4) Printed clear polypropylene
   (5) Structural packaging and branding for product

5. (1) Lindores Abbey, Scotland
   (2) Lindores Abbey Whisky
   (3) Limited Release Premium Whisky
   (4) Hand-crafted wooden box lined in sack cloth and containing an individually signed and numbered bottle of whisky (1/150)
   (5) Structural packaging and branding for product

4.

5.

# CHA-CHING DESIGN
### THAILAND

1  Marriott International
   Gold Card Membership Program
   Graphic and Packaing Design

2  The Barai, Hyatt Regency Hua Hin, Thailand
   Fisherman's Pants
   Graphic and Packaing Design

**It takes strong tools to build a strong business.**

The quality of your marketing materials can have a big impact on the success of your business. Done well, they can help you to promote sales, motivate customers and contribute to a strong brand image. Done poorly, they can do more harm than good.

At ChaChing Design we specialize in creating the visual tools companies need to achieve their marketing objectives. Whether you're interested in designing a website, developing a sales brochure, or establishing a new corporate identity, we can help you create the professional image you need to take your business to the next level.

Cha-Ching Design
33 Wang Dek Building 4
Soi Yasoop 1 5th Floor, 5C
Vibhavadi Rangsit Road, Jompol
Jutujuk, Bangkok 10900 Thailand
T+66 2 6177752
F+66 2 6177753
www.chachinggroup.com

3  Iddhi Hospitality, Thailand
   Natural Soap Packaging
   Illustration and Graphic Design

4  Iddhi Hospitality, Thailand
   Shampoo Conditioner Label
   Illustration and Graphic Design

# PROJECTGRAPHICS
KOSOVA

## project|GRAPHICS

Project Graphics is a leading creative factory that brings together a group of young and inspiring talents in the field of contemporary design, web design, package design and architecture.

It nurtures a diverse interactive creativity since 2002, when it was founded by Agon Çeta, its lead designer, as a response to the overwhelming demands over his illustrious straight lined designs that marked his generation.

The studio nourishes the concept of becoming one with the client, dousing the traditional perception of this bond and creating a new entity that strives to achieve perfection, both aesthetically and commercially. The greater the challenge, the stronger the ties become, as one side gives shape to the thoughts of the other in multi-layered creations that are both practical and beautiful.

www.projectgraphics.eu

projectGRAPHICS
Str. Ilaz Kodra, H.4 Nr.7a
Kroi i Bardhë, Prishtinë
Kosova
T +377 44 124 139
info@projectgraphics.org
www.projectgraphics.eu

1 RUGOVE, Kosova
RUGOVE Spring Water 0.5L
Brand Identity and OPP labels
Graphic Design

2   RUGOVE, Kosova
    RUGOVE Spring Water 24x0.25L
    Package
    Structural and Graphic Design

3   RUGOVE, Kosova
    RUGOVE Spring Water 0.75L
    Brand Identity and OPP labels
    Graphic Design

# WILLIAM FOX MUNROE, INC
## USA

William Fox Munroe has been developing graphic design for packaging and advertising for over 35 years. We opened in 1972 as a one-man, one-client design shop. Things grew from there, and in 1998 things began to change when employees Dan Forster, Tom Newmaster, and Steve Smith purchased the business and began to focus exclusively on package design and creating a more compelling experience for the consumer at point of sale.

In the years since they purchased the agency, the partners have put programs into place that have helped our agency take off.

In addition to winning numerous design and packaging awards, WFM was voted #16 in Fast Company Magazine's Fast 50 in 2005. This list highlights the 50 most innovative companies in the world.

What's our secret to all this success? Differentiation.

We're focused. That means we only do packaging and packaging support. No paid media. Just package design 24/7.

We're fast. When you live in the packaging world, you understand that it doesn't matter if it's good if it's just too late. Deadlines are critical.

We're user-friendly. No attitudes or egos here. We make sure you never feel like we're doing you a favor by completing your project on time and within budget.

And finally, it's much easier to be successful when you love what you do. We care about what we do and it shows. And we plan to be loving our work for many years to come.

Please visit: www.wfoxm.com

William Fox Munroe, Inc.
Shillington, PA 19607 USA

Phone: 800.344.2402

The Hershey Company / USA
Hershey's Kisses
Confectionery
William Fox Munroe, Inc.

Paperboard Packaging Council
Excellence Award

1
Murry's / USA
French Toast Sticks
Frozen Foods
William Fox Munroe, Inc.

2
Tom Sturgis Pretzels / USA
Artisan Pretzels
Snack Foods
William Fox Munroe, Inc.

3
CoolFactor™
Futuristic Packaging Concept
Confectionery
William Fox Munroe, Inc.

4
Simple Servings™ (patent pending)
Futuristic Packaging Concept
Breakfast Cereal
William Fox Munroe, Inc.

5
Wyeth Consumer Healthcare
Advil Day & Night
OTC Pain Reliever
William Fox Munroe, Inc.

6
Wyeth Consumer Healthcare
AdvilPM Liqui-Gels
OTC Pain Reliever
William Fox Munroe, Inc.

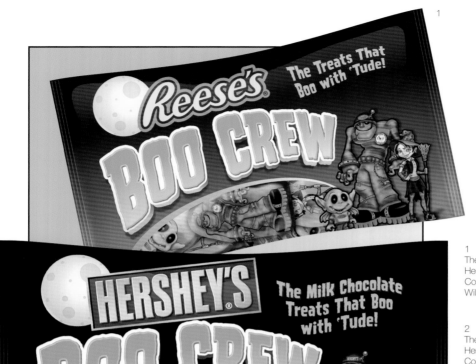

1
The Hershey Company / USA
Hershey's & Reese's Boo Crew
Confectionery
William Fox Munroe, Inc.

2
The Hershey Company / USA
Hershey's Milk Chocolate Hoppy Bunny
Confectionery
William Fox Munroe, Inc.

3
The Hershey Company / USA
Whoppers Sno-Balls
Confectionery
William Fox Munroe, Inc.

Whoppers Sno-Balls Awarded:
Excellence in Flexography
2008 Gold & Best of Show:
Graphic Design

4

4
Armstrong World Industries / USA
Bruce Hardwood & Laminate
Floor Cleaner
William Fox Munroe, Inc.

5
Woodstream / USA
Safer Brand 3-in-1 Garden Spray II
Lawn & Garden
William Fox Munroe, Inc.

5

6
Stoner / USA
Invisible Glass
Glass Cleaner
William Fox Munroe, Inc.

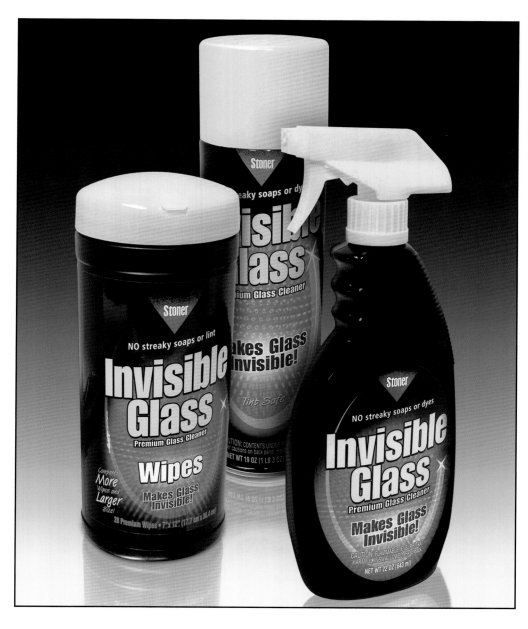

# BHZ DESIGN
## BRAZIL

Left to right: André Reinke, principal and design director; Carlos Pires, designer; Euler Silva, principal and design director; Stefan Wondraceck, designer.

BHZ Design is a studio which concentrates in building brands and managing creative solutions for packaging. It is composed of a team of talented designers, all who began in branding, and that have been active in the market for over 12 years.

Their working method combines all steps of the project, from identifying the needs, concept, creation, product positioning, and production supervision. The methodological approach searches for packaging solutions, providing excellence in the concept and result, offering client´s consumers products focused on a new consuming experience. Efficient packaging should provide brand equity, and consequently, an increase in market share.

The team has solid experience in legal matters and in market requirements, offering clients the support necessary for nationalizing and adapting projects to various markets, complying with globalized norms.

BHZ believes in the success of its projects, because they contribute effectively to the construction of the clients´ brand.

**www.bhzdesign.com.br**

BHZ Design
Rua Ramiro Barcelos 1215 | 401
90035-006 | Porto Alegre | RS
Brazil
T/F+55 51 3024 8030
bhz@bhzdesign.com.br
www.bhzdesign.com.br

1 Ecocitrus | Brazil
Organic Juice
Label Design

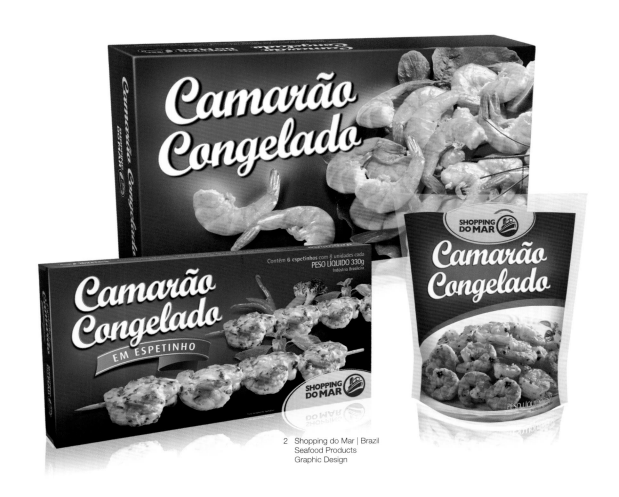

2  Shopping do Mar | Brazil
   Seafood Products
   Graphic Design

3  Harpla | Brazil
   Coffee Filters
   Structural Packaging & Graphic Design

4　O Bananeiro | Brazil
　Chocolate covered Banana Products
　Graphic Design

5　Naturale | Brazil
　Oatmeal Products
　Graphic Design

6　Allclub | Brazil
　Sports Marketing
　Structural Packaging & Graphic Design

# THE GRAIN
## AUSTRALIA

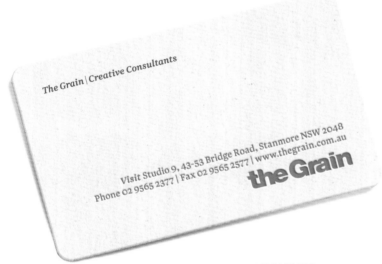

## ABOUT US

In only five years, the Grain has built an enviable portfolio by developing client relationships based on trust, collaboration and inspiring creativity.

We work on branding and packaging projects but specialise in creating value through ideas.

Our small team of designers and strategists are also parents, consumers, dreamers and citizens. As people, we celebrate the varied motivations of other people and use this way of working to unlock channels of communication, which help our clients connect with their audiences.

We like to say that we design from a consumer's perspective through a combination of passion and methodology.

We understand a brand is not built on intention alone but on consistent behaviour that reflects its values. That's why we try to do our bit through energy efficient office management as well as through our connections and mindset.

We influence our clients to exercise social responsibility by building sustainable brands because we know all profits are ultimately reliant on people and the planet.

We also actively manage a variety of pro bono client accounts including charity organisations and educational support for aspiring designers.

**The Grain** | Studio 9, 43-53 Bridge Road, Stanmore, NSW 2048 | Australia
T +61 02 9565 2377 | F +61 02 9565 2577
weare@thegrain.com.au | **www.thegrain.com.au**

Lowan Whole Foods, Australia
Eden Museli
Brand development and
packaging design

**CASE STUDY: GREEN'S BAKING**

Green's Baking sells an emotional product in a fast paced and hectic grocery environment. So on this project we evolved the brand and revolutionised the packaging to better target consumers at the heart of their emotional connection with each occasion that is relevant to the specific product variants.

This repositioning exercise was driven predominantly by the packaging design, however our role extended to the development of print advertising, category theatre, trade communications and presentations.

The working process started in collaboration with the client and a research company as we sought to develop buyer segments. The objective was to inject more relevance for the consumer as well as add value, convenience and inspiration.

The designs we created communicate the identified occasions through icons, colours, photography and styling, while maintaining a strong brand link in layout.

A new hatbox pack format was developed for larger kits. It introduced a handle for portability, lid for accessibility and tamper sealing for quality assurance.

Texture was added to all the packs with the logo embossed and gold foiling on premium chocolate variants.

The Green's brand was rejuvenated with a new logo, which introduces a softer colour tone and holding device, giving it a more contemporary and dynamic appearance without compromising on brand blocking and shelf impact.

Green's, Australia
Green's Baking
Brand development, structural design, graphic design and photography

Lowan Whole Foods, Australia
Lowan Kid's Cereal
Brand development
and packaging design

# FiF DESIGN house
## THAILAND

# FiF DESIGN house
### Branding & Identity Design Specialist

**Getting to know us!**

FiF DESIGN house is a Branding and Identity Design Consultancy based in Bangkok, Thailand. We help our clients build brands and businesses by delivering new experience in strategies to design executions. Our multidisciplinary team offers compelling creative and extraordinary solutions, ranging from branding issues, design research, product design, retail environment design, packaging design, and communication design.

**What we believe!**

Design is the great story to make better difference to business and people life.

www.fifdesign.co.th

1   AXII
    Cosmetic
    PP bottle
    Bottle Structure & Graphic Design

2   Nestle Bear Brand
    Nestle (Thai)
    Powder Milk Beverage
    Box
    Graphic Design

**FiF DESIGN house**
200/1 Phibulwattana 1/1,
Rama 6 Rd., Samsaen-nai,
Phayathai, Bangkok 10400,
Thailand
T: 66+2278+2538
F: 66+2615+7671
Email: info@fifdesign.co.th
www.fifdesign.co.th

3   Milo | Thailand
Nestle (Thai)
Chocolate Malt Mixed Beverage
Foil Sachet
Graphic Design

# BUTTERFIRE
## THAILAND

Arun Sittisuporn & Rodjanai Phanpruk, Co-design director

Butterfire is Packaging and Graphic design consultancy firm based in Bangkok, Thailand. Provides the practical solutions to client ranging from brand identity, packaging design and character design.

Our service not only produce aesthetic pleasure design, but also translate the marketing concept into creative and communicable design because we believe the good design must represent the identity of product and distinctive amongst the fierce competition in the present day.

Our corporate philosophy is simple : "Flying together". We always consider clients as our partners, their success is our passion.

**www.butterfire.com**

Butterfire Co.,Ltd.
118 Soi Romklao 36, Romklao Rd.,
Latkrabang, Bangkok, 10520, Thailand
T+66 2737 5874, F+66 2737 5875
info@butterfire.com
www.butterfire.com

1 Doi Nam Sub Co.,Ltd. | Thailand
Doi Nam Sub
Herbal Compress
Cardboard/4-color
Structural packaging and graphic design

2 Pro-dairy Co.,Ltd. | Thailand
Zezame
**Roasted black sesame**
Cardboard/4-color
Structural packaging and graphic design

3   Srithai Superware PCL. | Thailand
    SNatur my soup
    Instant soup
    Cardboard/4-color
    Structural packaging and graphic design

5   Samuthanakom Co.,Ltd. | Thailand
    Choa-Lay
    Seafood snack
    Laminate plastic/7-color
    Logo and graphic design

6   VPP Progressive Co.,Ltd. | Thailand
    VPP
    Ground coffee
    Aluminium foil/7-color
    Graphic design

7  Elements Fengshui | Dubai
Elements Fengshui
Aroma incense
Cardboard/2-color with Silver Hotstamp
Structural packaging and graphic design

# CILINDRINA
## ITALY

Cilindrina Creative Consulting was founded in 1997 by Georgia Matteini Palmerini, in the role of art director and illustrator. Cilindrina Creative Consulting creates drawings, graphic animation, editing project, packaging design, gadget and corporate identity for public and private clients.

All projects are create with love and passion with a touch of humour and irony. That's the Cilindrina style.

Cilindrina
Via dei Mille 34
47900 Rimini
Italy
T/F+39 0541 788122
info@cilindrina.it
www.cilindrina.it

1 Garagnani Ricerche Cosmetiche | Italy
Thermae Latinae
Body health care
PET and glass bottles / Natural paper labels
Structural packaging & label design

293

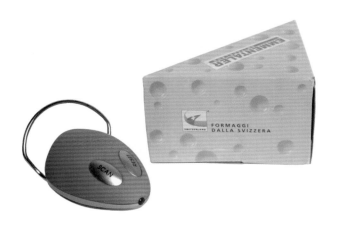

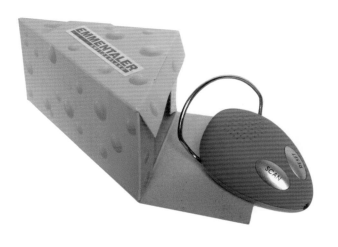

2   La Table & Bijoux | Italy
    Emmentaler - Switzerland
    Gadget radiomouse
    Cardboard box
    Structural packaging & box graphic design

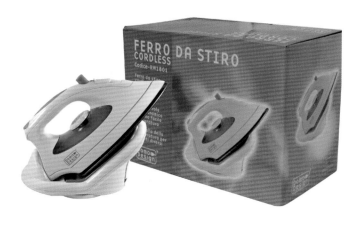

3　La Table & Bijoux | Italy
　　Domodesign
　　Suppliers
　　Cardboard box
　　Box graphic design

# BOUTIQUE GRAFICA
### CHILE

BG Team:
Partners, Paola Leyton, Paola Ferruz, Ana Maria Villanueva
Project Manajer, Veronica Leyton
Designers, Mariana Leyton, Carlos Peñailillo

The underlying philosophy of "BOUTIQUE" is to create exclusive designs that are specifically produced to meet individual requirements. We believe a good design provides "added value" to any product or service. It is our aim to provide clients with a service based on reliability, rigorous quality control and design excellence.

We Are Creative !
We live, think and are passionate about Design!

What do we think?:: Design" gives the face and the first impression of every Company and Product.
How do we work?:: By listening, observing, investigating, analysing, questioning, proposing, correcting and learning. A good communication with the Client is the key to the best result.

www.boutiquegrafica.cl

BOUTIQUE GRAFICA
5 Oriente 457,
Viña del Mar, Chile
T/F+56 32 2111486
contacto@boutiquegrafica.cl
www.boutiquegrafica.cl

**Coheso, USA**
**CalorieSmart** package design
Handheld Calorie counter
Graphic Design

Mood Menders, USA
Eye Dews – display package
Graphic Design

Icebreaker - USA
Icebreaker Cologne - Brand and package design
Graphic Design

# MICHAEL FUSCO DESIGN
## USA

Humans (left to right): Emma Straub, assistant designer and conceptual genius; Michael Fusco, art director and owner;
Felines (left to right): Gravy Boat Fusco, Killer Guerilla Straub.

## michael fusco design.

Michael Fusco has lived in climates as varied as Jupiter, Florida, Brooklyn, New York, and Madison, Wisconsin. After working in publishing for many years, Michael decided to strike out on his own, which has given him the opportunity to design more book jackets, record covers, and logos than one could comfortably shake a stick at. Michael's work has been celebrated in Print Magazine's Regional Design Annual, and he has won awards from the AIGA and the National Calendar Awards.

**www.michaelfuscodesign.com**

Michael Fusco Design
917.974.3032 (Phone)
michael@michaelfuscodesign.com
www.michaelfuscodesign.com

1  Wheelhouse Pickles | Brooklyn, NY
   Packaged pickled vegatables
   Logo & label design

# THE O GROUP
USA

We're The O Group. The Strategy and Design Agency for Luxury Brands. Our work is the result of creative thinking on every level. We want to understand it all—your brand, your audience, your goals. Our expertise lies in crafting thoughtful solutions, flexing our creative muscles to produce ideas and designs that connect with the luxury consumer.

259 W 30 STREET / NYC 10001
T/ 212.398.0100  F/ 212.398.9191
WWW.OGROUP.NET

1 Lacoste, USA
  Lacoste
  Holiday Packaging
  Polypropylene with white silk screen
  The O Group

2 Ruby et Violette, USA
  Ruby et Violette
  Gift Packaging
  Coated paper with spot UV varnish,
    clear foil stamp and silk ribbon
  The O Group

3 Robert Marc, USA
  Robert Marc
  Gift Box & Bag
  Coated paper with matte
    lamination and grosgrain ribbon
  The O Group

# WPA PINFOLD
## LEEDS/UK

**wpa**PINFOLD

WPA Pinfold has worked with niche and multinational clients from a wide range of industries, designing world class brands and motivating packaging for over 25 years. As designers of brands and communications, our role is to interpret clients' needs, clarify the message and build brand power.

Everyone is constantly bombarded with information. We cut through the visual clutter to help focus on the key concepts that support the brand. We know what works – and how to use design to deliver success. Our ability to innovate successfully comes with experience – of markets, customer behaviour and the practical limitations of production techniques.

We believe that the key to managing and protecting brands is understanding exactly where their equity comes from. At WPA Pinfold, we do this by stripping a brand down to its bare essentials – to discover the concepts and icons that help build market loyalty.

Brands exist in people's hearts and minds, yet their power lies in tangible features – such as colours, shapes, words or images. Working with a myriad of shapes, sizes and materials, we continue to retain the same passion and respect for this most challenging of art forms.

**Client:** Rocket Fuel Drinks Co, Scotland
**Product:** Rocket Fuel Vodka

**Client:** Delvaux, England
**Product:** ProCoco, Probiotic Chocolate

**wpa**PINFOLD

Client: T&R Theakston, England
Product: Old Peculier, Seasonal and Smooth Dark Beer

Client: G Pharma, England
Product: Prescription Only Pharmaceuticals

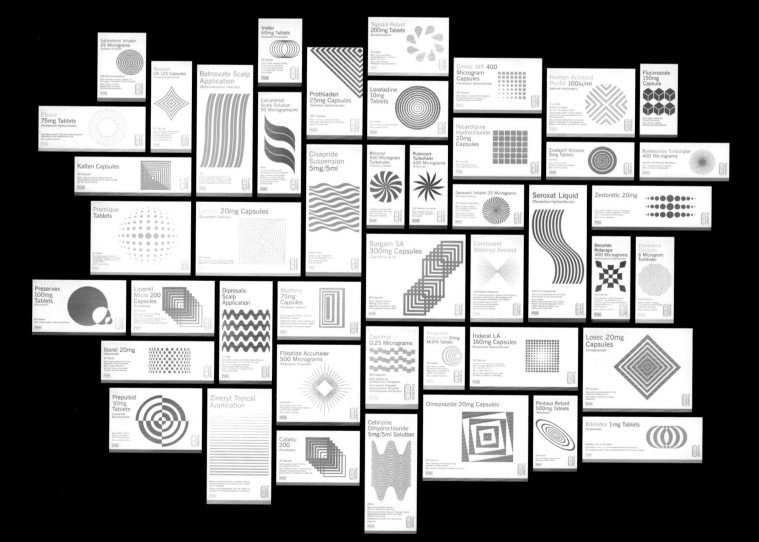

# MIRAN
## BELARUS

1  Five Stars
   For Cosmetica XXI company.
   Shampoo bottles for men series.

**Miran**
Republic of Belarus
Minsk
Kalinovskogo 74-1-77
220114
**Igor Solovyov**
http://www.solovyovdesign.com/
http://www.miran-bel.com/

2 Yogurt dessert
  For April company
  Shampoo bottles.

3 Sirius and Zodiac
  For Cosmetica XXI company.
  Concept packaging for new line premium cosmetics
  with medicine effects. Shampoo bottles.

4 Rosa
  For Miran company
  Shampoo bottles.

5 Olives
  For Cosmetica XXI company.
  Shampoo bottle and jar

6 Kuvshin
  For Cosmetica XXI company.
  Shampoo bottle and jar.

# BLOK DESIGN
## MEXICO

A design studio and a space for ideas, BLØK is at once idealistic and intuitive. It is also international in character and scope, attracting talent from throughout Mexico and around the world. Together, we collaborate on initiatives that blend cultural awareness, a love of art and a fundamental humanity to advance society and business alike.

www.blokdesign.com

1 Aspenware, Canada
   Wün, (wooden cutlery)
   An identity for the new line of wooden cutlery for Aspenware, a new company based in British Columbia elevates the product above the "natural food" niche.

2a Caban, Canada
In developing the identity for Caban, we needed to reflect the interrelation of classic and modern products from the branding to the different packaging applications. There is a simple, no-nonsense yet stylistic approach which sees it's way through the entire system.

2b  Caban, Canada

3   ICC, Canada
    ICC, a chemical company positions its organic fertilizer, plant food and compost products as the future — and chemically-derived products as the past. Packaging expresses the essence of a company committed to the environment.

# chatchan.com

たとえ英会話が下手でも、ネイティブ（アメリカ、イギリス、カナダ、オーストラリア）の人達と勇気を出して直接 話をしてみましよう！
200人を越える英語圏の人達が、あなたの英会話力を向上させるよう、いつでも待機しています。

# chatchan.com

インターネット回線で、通話料は無料。現地の学生を中心として、主婦、学校の先生、サラリーマンの有志の方々が、あなたと
一対一で日常の楽しい会話をしてくれます。費用は有志の人達へのアルバイト代のみです。

詳細はチャットちゃんサイトをご覧下さい：    International Creators' Organization (ICO)
http://chatchan.com ■    ■ Contact to : ICOHQ. icohq@info.email.ne.jp

# Welcome To Chat-Chan

International Creators' Organization is proud to present Chat-Chan, a revolutionary new program to help Japanese students learn to speak conversational English. Chat-Chan is a "person to person" interactive learning and cultural exchange program between ordinary people.

ICO has been a respected leader in the world of art and design for the past 30 years. In addition to our publication of International art books, ICO is also well known as a meeting place for outstanding creators and distinctive clients worldwide. In today's world the art of verbal communication is rapidly becoming equally as important as the art of visual communication. Therefore, ICO is now expanding the fields of art we represent to include the "Art Of Conversation". In order to fill this need ICO has created Chat-Chan.

At ICO our dream for the future is to help work toward a peaceful world, with one language, where people of all countries will have a better understanding of each other. Through our Chat-Chan program we hope to play a significant role in helping to make this dream become a reality.

Thanks to the enthusiastic response from over 170 teachers from America, Australia, Canada, and the UK, we have successfully achieved the first stage of our Chat-Chan project. Please meet these teachers on the Japanese page of our Chat-Chan site.
<http://chatchan.com/chatchan_new/index.html>
By clicking on the teacher's image you will be able to jump to the teacher's profile data.

ICO is currently introducing Chat-Chan to the Japanese market. As soon as we have successfully completed this second phase of our program Chat-Chan will begin.

We cordially invite you to join us by participating in Chat-Chan. We believe that person-to-person, and bit by bit, we can achieve better global understanding between people of all countries, and all languages, through this revolutionary new program.
To learn more about Chat-Chan please view our Chat-Chan site at:
<http://chatchan.com/>http://chatchan.com/

**For more information go to :**
ICO Headquarters : Norio Mochizuki <icohq@info.email.ne.jp>
ICO Japan : Satoru Shiraishi <c-wave@gaea.ocn.ne.jp>
ICO America : Robert Morris <icoamerica@cox.net>
ICO America / Kansas : Jo Ssickbert <Josanartist@aol.com>

# INDEX

This index page synchronizes with each designer's home page via ICO's website. Please click the following ICO site/ so that you can visit their fantastic studio and outstanding works worldwide.

STEP-1  http://www.1worldart.com/aa/jb/first_page_package_d/package_cover.html
STEP-2  http://www.1worldart.com/aa/jb/first_page_package_d/serch_by_designer/serch_by_designer.html

| | | | |
|---|---|---|---|
| | ALOOF DESIGN | UK | 40-43 |
| | ANNETTE SCARFE DESIGN | SPAIN | 34-39 |
| | ARBOL | ARGENTINA | 182-185 |
| | AUSTON DESIGN GROUP | USA | 44-51 |
| | BHZ DESIGN | BRAZIL | 274-277 |
| | BLACKANDGOLD | FRANCE | 92-95 |
| | BLOCK DESIGN | MEXICO | 310-313 |
| | BOUTIQUE GRAFICA | CHILE | 296-299 |
| | BOUZON I COMUNICACION Y DISENO | ARGENTINA | 220-225 |
| | BRITTON DESIGN | USA | 170-175 |
| | BRUNAZZI&ASSOCIATI | ITALY | 144-149 |
| | BUTTERFIRE | THAILAND | 286-291 |
| | CATO PARTNERS | AUSTRALIA | 14-19 |
| | CHA-CHING GROUP Co.,Ltd | THAILAND | 264-265 |
| | CHEN DESIGN ASSOCIATES | USA | 66-73 |
| | CILINDRINA | ITALY | 292-295 |

このインデックスページは、ICO のウェブサイトで各デザイナーのホームページにリンクしています。
以下の ICO のホームページから、それぞれの作家のイメージ・アイコンをクリックしてみて下さい。
世界中のファンタスティックな彼らのスタジオや、素晴らしい作品の数々を訪ねることができます。

STEP-1 http://www.1worldart.com/aa/jb/first_page_package_d/package_cover.html
STEP-2 http://www.1worldart.com/aa/jb/first_page_package_d/serch_by_designer/serch_by_designer.html

| Designer | Country | Pages |
|---|---|---|
| CINQ DIRECTIONS INC. | JAPAN | 216-219 |
| DENEY DESIGN | TURKEY | 198-201 |
| DEPOT WPF | RUSSIA | 230-237 |
| DESIGN FORCE | JAPAN | 134-137 |
| DESIGNAFFAIRS | GERMANY | 64-65 |
| DEZIRO | JAPAN | 86-91 |
| DIL BRANDS | BRAZIL | 154-161 |
| DOLHEM DESIGN | SWEDEN | 100-105 |
| EICHE,OEHJNE DESIGN | GERMANY | 192-197 |
| ESTUDIO IUVARO | ARGENTINA | 20-25 |
| FIF DESIGN HOUSE | THAILAND | 284-285 |
| JOAO MACHADO | PORTUGAL | 52-55 |
| KHDESIGN GMBH | GERMANY | 128-133 |
| KIRK WYLAM DESIGN | UK | 242-243 |
| LEWIS MOBERLY | UK | 8-13 |
| LITTLE BIG BRANDS | USA | 150-153 |

# INDEX

| | | |
|---|---|---|
| LLOYD GREY DESIGN | AUSTRALIA | 96-99 |
| MARTA ROURICH | SPAIN | 80-85 |
| MICHAEL FUSCO DESIGN | USA | 300-301 |
| MILK | GREECE | 254-257 |
| MILLER CREATIVE LLC | USA | 138-143 |
| MINIRA CREATIVE | JAPAN | 206-209 |
| MIRAN | BELARUS | 308-309 |
| NARROW HOUSE | SPAIN | 210-215 |
| NO.PARKING | ITALY | 106-109 |
| PROAD IDENTITY | TAIWAN | 56-63 |
| PROJECTGRAPHICS | KOSOVA | 266-267 |
| QUON/DESIGNATION | USA | 176-181 |
| SABINGRAFIK,INC | USA | 248-253 |
| SIO DESIGN | JAPAN | 110-115 |
| STUDIO GT&P | ITALY | 124-127 |
| STUDIO360 | SLOVENIA,EU | 186-187 |

ICO HQ. / Norio Mochizuki
< ico-nori@info.email.ne.jp >
ICO AMERICA California / Robert Morris
< ICOAmerica@cox.net >
ICO AMERICA Kansas / Jo Sickbert
< JOSICKBERT@aol.com >
ICO JAPAN / Satoru Shiraishi
< c-wave@gaea.ocn.ne.jp >

| | | |
|---|---|---|
| SYLVIE DE FRANCE, DESIGNER | FRANCE | 238-241 |
| THE CREATIVE METHOD | AUSTRALIA | 188-191 |
| THE GRAIN | AUSTRALIA | 278-283 |
| THE KITCHEN COLLABORATIVE | USA | 226-229 |
| THE O GROUP | USA | 302-303 |
| THEORY ASSOCIATES | USA | 244-247 |
| TRIDIMAGE | ARGENTINA | 162-169 |
| TRIPLE 888 STUDIOS | AUSTRALIA | 202-205 |
| TUCKER CREATIVE | AUSTRALIA | 116-123 |
| UP CREATIVE | TAIWAN | 74-79 |
| VIE DESIGN STUDIO | INDONESIA | 258-261 |
| VILLEGERSUMMERSDESIGN | UK | 26-33 |
| WILLIAM FOX MUNROE, INC | USA | 268-273 |
| WPA PINFOLD | UK | 304-307 |
| ZERO DESIGN LIMITED | SCOTLAND | 262-263 |

# Epilogue

Once again ICO is proud to present to you outstanding representations of the latest trends in package design, from many talented designers worldwide.

As we study and analyze the ever changing trends of today's global marketplace we can see that the package design industry is constantly evolving to meet the needs of the consumer. Every day, continuously, designers are faced with the challenge of creating innovative and inspired solutions for these ever changing trends and needs.

Needless to say, trends in package design have changed dramatically over the years. We at ICO have always followed these trends closely. ICO's first noAH publication was in March of 1985. Our goal was to share with you the latest trends and talent in the field of International package design. Since then 24 years have passed. We have periodically compiled more books in our noAH series, always keeping our goal of bringing you the best of cutting edge package design, created by top International designers.

noAH-8, ICO's last package design annual, was published just a short time ago, in January of 2008. However, already there have been many changes in trends of the global marketplace. Of course the most pressing change has been the downturn of the global economy. Every country around the world is faced with difficult economic circumstances, presenting many challenges to all businesses, including the package design industry. Another change has been the increasing awareness of environmental issues. The whole world has become more conscious of the need to pursue a more eco-friendly lifestyle, including the products we consume, and even the packaging of these products. There is a worldwide trend to "go green". More than ever, designers are challenged to create visionary and innovative new design solutions to fit these real world needs.

Although trends of the global marketplace are continuously changing, there will always be a need for the visual communication of good package design. The world also continues to need the creative talents of outstanding designers in all countries, to fulfill these needs.

We at ICO wish to offer our sincere thanks to the many talented designers worldwide who have generously shared their outstanding creative package design works with us for noAH-9. As we have the opportunity to view their excellent work, we have a strong feeling of confidence for a bright future in the package design industry. The passion and talents of these gifted designers hold great promise for a future filled with unlimited possibilities. Here at ICO we plan to follow this dream of a bright future in upcoming noAH publications.

All ICO Editorial Group

## ICO noAH-9 Compilation Project

ICO's editorial project has no specific office. It is a space information project, connected from vast points, with ICO HQ. at the center.
The points of Chigasaki, Sapporo, Kansas, Los Angeles, Milano and Paris are all connected on the Internet. Information flows freely from each point every day.
The crystallization of ICO's editorial project this time is the publication of noAH-9, an International package book. noAH-9 is a continuation of ICO's world renowned noAH series.

Norio Mochizuki, · · ICO HQ.
Jo Sickbert, · · ICO Kansas
Robert Morris, · · ICO Los Angels
Yumiko & Toshie Mochizuki, · · Editorial staff
Satoru Shiraishi, · · ICO Japan
Kazuhito Mochizuki, · · Technical staff

Norio & Yumiko Mochizuki

Jo & Wally Sickbert

Robbert & Fath Morris

Kazuhito    Toshie M.

Satoru Shiraishi

---

■ **Title**
World Package Design - noAH -9
■ **Released**
October. 2009
■ **Art Supervision**
Yumiko Mochizuki, Jo Sickbert, Robert Morris, Toshie Mochizuki, Colette Coltte, Satoru Shiraish
■ **Cover Design** Norio & Kazuhito Mochizuki Lewis Moberly (UK.)
■ **Cover images**
Lewis Moberly (UK), Prologue: Aloof design (UK)
■ **Editorial Design Layout**
Yumiko & Toshie Mochizuki,
■ **Translation**
Jo & Wally Sickbert, Robert morris
■ **Technical support**
Kazuhito Mochizuki, Robert Draper
■ **Publishing House**
ICO CO., LTD. Publishing House,
ICO ( International Creators' Organization )
Post Code 253-0051 13-14 Wakamatsu-cho, Chigasaki, Kanagawa, JAPAN
t : 03-3292-7601   f : 03-3292-7602
url : http://www.1worldart.com/
■ **ICO site**
http://www.1worldart.com/
■ **noAH INDEX site**
http://www.1worldart.com/aa/jb/first_page_package_d/package_cover.html
■ **Publisher**
Norio Mochizuki

------------------------

**ICO Distribution Activities :**
ICO HQ.
13-14 Wakamatsu-cho, Chigasaki, Kanagawa, JAPAN
Postal code : 253-0051
t    : 81(Japan) (0)467-87-2110
f    : 81(Japan) (0)467-86-8944
e    : <ico-nori@info.email.ne.jp>

**Distributor :** AZUR corporation
1-44-8 Jinbou-cho, Kanda, Chiyodaku, Tokyo
t : 03-3292-7601   f : 03-3292-7602

---

タイトル / World Package Design "noAH - 9"
発行日 / 2009 年 10 月
表紙デザイン / 望月 紀男・かずひと
編集制作 / 望月 由美子・俊江、ジョー・シックバート、ロバート・モーリス、望月 かずひと、白石 智
本文レイアウト・デザイン構成 / 望月 由美子・望月 俊江
翻訳 / ジョー & ウォーリィ・シックバート
発行所 / (株) ICO
〒 253-0051 神奈川県茅ヶ崎市若松町 13-14
t : 050-3424-5072   f : 0467-86-8944
url : http://www.1worldart.com/
noAH INDEX サイト：
http://www.1worldart.com/aa/jb/first_page_package_d/package_cover.html

発行人 / 望月 紀男

ISBN 978-4-931154-30-8C3072